LIFE'S BEGINNINGS

*Our Life Before Birth
Helps Us Trace Our Roots*

W. J. Howard

Illustrated by the author

Coast Publishing

Coos Bay, Oregon

Cover photo by Scott Blackman

Published by
Coast Publishing
P.O. Box 3399
Coos Bay, OR 97420

Library of Congress Catalog Card Number: 90-83709
ISBN 0-9627341-7-9

Printed in the United States of America

To that marvelous speck
we all were once

CONTENTS

PREFACE

My purpose in writing this book is to help people understand a fascinating story that concerns all of us. "Fascinating" sounds like advertising hype but, as you'll soon discover, we're now finding answers to some of the questions that have puzzled human beings since they first learned to speak: Where did we come from? What is the origin of life? Not many people are aware of how much we've learned about these things; some may be able to see a tree here and there, but they're missing the forest. Even those who live in the fast lane are way behind the times. The purpose of this small book is to help bring people up to date.

There are other books on these subjects. Why add yet another, particularly a thin book with lots of pictures? Let me explain.

Yes, there are other books, but how many people take the time to read them? The information is scattered, and books that are at all technical can be intimidating; thick, wordy tomes are not the best way to get a few basic ideas across to casual readers. They remind us of the little girl who returned a book to the librarian. Asked how she liked it, she said: "It told me more than I wanted to know about penguins."

In planning this book I realized that the story of our origins will just be dismissed as incredible if it's not understood. To help you see the "big picture" and not get lost in details, I've made it easy to skim through for the basic ideas. A typical page has a heading and picture expressing the main idea, as shown in the following diagram:

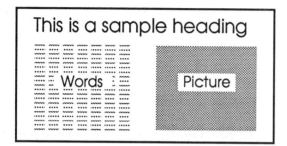

Skim through looking at the headings and pictures. Then if you want more details, read the words.

Why the picture? The ideas in this book are unfamiliar, so we need all the help we can get. We've all heard the saying "A picture is worth a thousand words." But what we really need is both words *and* pictures. When you're reading, only the left half of your brain is working. Pictures bring the right half into play. So with both words and pictures, both halves are chugging along—your whole brain, not just part of it, is working for you.

Here's how the book is arranged: We'll begin close to home—our growth from tiny speck to human being. An absorbing tale in its own right, but we're just getting started. Then we'll trace our origin back to the dawn of life itself, in the process finding amazing similarities between our own life before birth and earliest life on earth. More than similarities, really; we all carry in our bodies telltale signs of our ancient origins and kinship with other life. But we won't stop even there. Before we're through, we'll work our way back to the origin of the universe—the beginning of time.

I don't pretend to be an expert in these areas. My contribution is to put the ideas together in a way that highlights the essence of an astonishing story. I also act as an interpreter. As a nonexpert I can appreciate the confusion others may feel, and help them over the hump. We need this kind of help in many areas (ever had trouble with a legal document? A financial statement, insurance policy, assembly instructions, math book?). Using a few basic principles (which I've summarized in an article in *BYTE* magazine, March 1981), I've tried to fill this need in various areas, but none as awesome as this one.

There are bound to be omissions and oversimplifications in a small book dealing with such huge questions. Many of the gaps are filled by the excellent references listed in the Bibliography. I think that once you see the big picture, you'll *want* to read some of these books and articles. If you do become interested enough to follow up on some of the ideas discussed here, this book will have done its job.

Which reminds me to thank those responsible for the references listed. But I don't want to give the impression that they agree with everything in this book or endorse my way of saying it. Any errors are of course my responsibility.

My thanks also to Drs. Kay Inaba and John Lewis for encouragement on the first draft; to family, friends and others for their kind comments (a few of which appear on the back cover); to Professor Douglas Hofstadter for pointing out an inaccurate statement; to Mary Barker for some editing tips and help with cover layout; to Sally Kostal for editing (and for detecting a glaring error); and to photographer Scott Blackman for a striking cover photo.

What we'll be talking about touches on religion, the abortion controversy, racial attitudes, and environmental questions, but these issues will be avoided. It's best to leave them to each individual to work out. Certainly people are more likely to make the right decisions in these (or any other) areas if they are adequately informed.

W. J. Howard

1

FROM TINY SPECK TO HUMAN BEING

Anyone insensitive to the wonders of life should be required to consider what was involved in creating him from a newly fertilized human egg, a tenth of a millimeter in diameter.

— Nigel Calder

Each of us begins life as a tiny speck, . . .

Aₛₖ ₛₒₘₑₒₙₑ "When did your life begin?" and chances are he or she will give you a funny look and say "On the day I was born, of course." He might even add something like "What a silly question!" If reminded that he was also alive the day before he was born, and even the day before that, he might start to get confused.

We tend to overlook the fact that, though we celebrate it every year, our birthday was not the real beginning of our life. That great adventure began—probably unnoticed—nine months earlier at conception, when our father's sperm united with our mother's egg cell. (It's for this reason that a Chinese tradition says a baby is a year old at birth.)

What did you look like then, at the real beginning? Actually, it's easy to draw an accurate portrait (see figure): When first starting out in life, you were a tiny speck just barely visible to the naked eye.

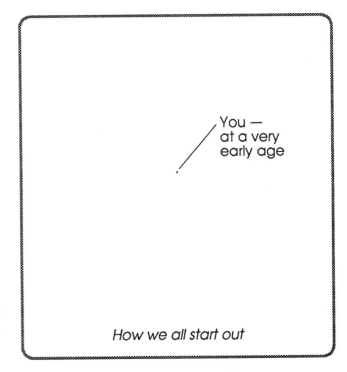

You —
at a very
early age

How we all start out

which eventually produces an adult human being.

Eventually that speck grows into a baby and then an adult human being—a human being who talks, thinks, and may even ask questions about life.

This transformation of a tiny speck—a single cell, a fertilized egg—into an adult with trillions of cells is so much a part of our daily experience that we tend to forget that this remarkable process is going on. What's happening here? How does one cell become trillions? By repeatedly dividing. Divisions of the original cell produce, in nine months, a baby with about two trillion (2,000,000,000,000) cells, and in about 20 years, an adult with some 60 trillion cells.

These cells are what your body consists of, what you *are*: skin, hair, fingernails, bones, muscles, nerves, blood—everything. How can one cell become all these different things? A good question, which we'll have to postpone for a moment.

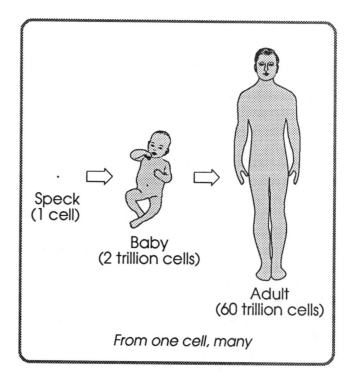

Speck
(1 cell)

Baby
(2 trillion cells)

Adult
(60 trillion cells)

From one cell, many

Growth begins as the cell starts to divide . . .

It all began when your father's sperm, after a long trip, met your mother's egg in one of her two fallopian tubes, which lead from ovaries to womb (uterus). The very tiny sperm was one of millions; the much larger egg was a single one, normally released from one of the mother's ovaries every 28 days. An initial stock of nutrients in the egg is what made it so "large" compared to other cells in your body; even so, the result of the union of sperm and egg is a barely visible speck.

Soon the speck starts to divide, as it begins its rather perilous trip down the tube. The first division is into two identical cells, which cling together (if they separate now, they'll become identical twins). Then each of these cells divides again, producing four cells. The divisions continue: 8 cells, 16, 32, and so on. In less than a week, if all goes well, what is now a ball of a few dozen cells arrives in the womb and attaches to the side.

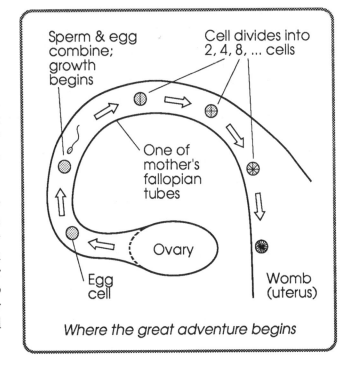

Sperm & egg combine; growth begins

Cell divides into 2, 4, 8, ... cells

One of mother's fallopian tubes

Ovary

Egg cell

Womb (uterus)

Where the great adventure begins

and continues over nine months to create . . .

The mother's womb provides a haven for the growing ball of dividing cells, surrounding it with a fluid-filled sac that cushions it from shocks. Food—which is necessary after what's in the original egg cell is gone—passes from the mother through an umbilical cord, an all-purpose tube that also channels oxygen from the mother and waste in the other direction. The umbilical does not go directly into the mother's body. It's connected to an absorbing surface—the placenta—created by the growing body of cells that's embedded in the womb. There are no direct connections between the mother and her offspring, and no blood is in common between them (however, substances in the blood can be exchanged).

The new life is called an *embryo* until it's eight weeks old. By then, though still tiny—only about 1 1/2 inches long—it has all its organs and other parts that will be in the adult human being. Its sex, though determined at conception, can now be observed for the first time. From eight weeks until birth it's called a *fetus*.

Growth via cell division doesn't stop with birth but continues until we're adults. The cells don't all divide at the same rate, and some are sloughed off in the growing process. (Each of us sheds many millions of cells every day of our lives.) Surprisingly, only about 46 divisions—as you can discover for yourself by continuing the sequence 1, 2, 4, 8, 16, ..., each number being double the previous one—are sufficient to transform the original speck into the 60 trillion or so that eventually form an adult human being.

Long before we became adults, we all went through the magic metamorphosis shown here. We were too "young" to know what was happening then, and can't remember now. What *was* going on?

a new baby.

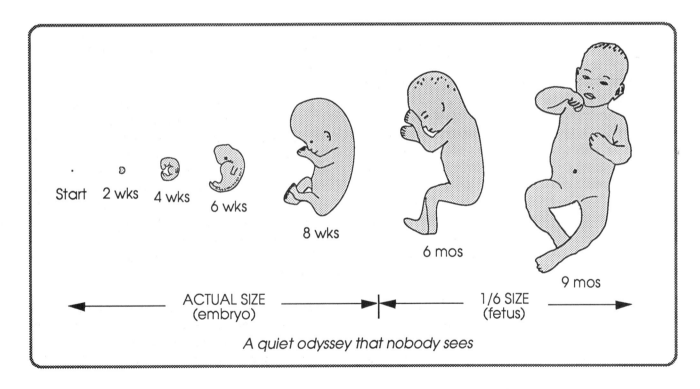

Start   2 wks   4 wks

6 wks

8 wks

6 mos

9 mos

ACTUAL SIZE
(embryo)

1/6 SIZE
(fetus)

A quiet odyssey that nobody sees

What was in that speck that caused this remarkable growth?

This question had for many years intrigued scientists, who were aware that there is more to that speck than meets the eye—the unaided eye, that is. Called a "cell" by Robert Hooke in the 17th century, the speck was pictured as a rigid honeycomb structure full of some strange life stuff. This was what he could see through his primitive magnifying lenses, enhanced by a fertile imagination. But he could not have imagined the beehive of activity within that we now routinely examine in detail. Not even Charles Darwin, who in the 19th century called the speck "perhaps the most wonderful object in nature," could have had any idea of the astonishing and unfamiliar world concealed within.

We have learned a great deal about this world, particularly in the latter half of this 20th century. Now that we know more about it, just what sort of a world is it?

More to it than meets the eye

At the heart of the cell: chromosomes and genes.

Most of the material in the egg cell from your mother was inert, nonliving food. (As mentioned earlier, this is what made it larger than typical body cells.) The "alive" portion of the egg cell was its tiny nucleus, which contained vital threadlike objects called *chromosomes.* The single sperm from your father contained chromosomes and little else. When the sperm entered the egg, its chromosomes combined with those of your mother in the nucleus.

Inside the chromosomes were *genes*—miraculous particles that direct your growth, establish your sex as male or female, and help determine what you'll be like as a human being.

How is it that these obscure bits can wield such a vast influence? Let's see how this mystery is gradually unfolding.

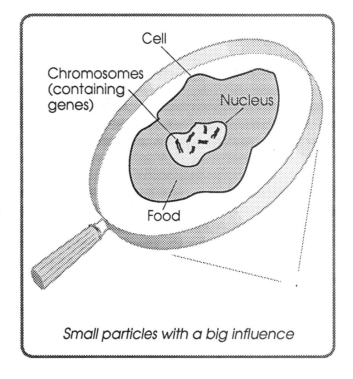

Small particles with a big influence

When a cell divides, two identical cells form, . . .

In the original speck (cell) that you once were, chromosomes occur in pairs: 23 pairs in all, for a total of 46 chromosomes. One of each pair comes from your mother and one from your father, so you get 23 from each parent. The 23 pairs differ somewhat from each other in shape and size, and do different things. When the cell divides, the chromosomes copy themselves, so that each new cell also has 46 chromosomes (23 pairs).

This continues with each cell division. In the end, each of your trillions of *body* cells—which need to be distinguished from your *germ* cells, the sperm or eggs—has all of your chromosomes. Therefore any of your body cells has the information to produce a complete you, a clone. Like a fingerprint, a single cell in, say, a hair identifies you—a fact now used by police in solving crimes.

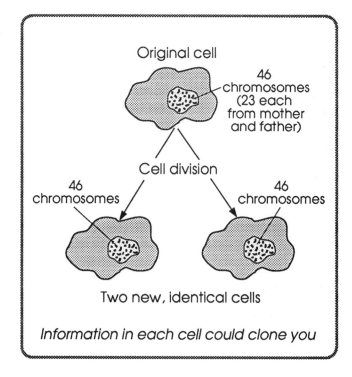

Original cell

46 chromosomes (23 each from mother and father)

Cell division

46 chromosomes

46 chromosomes

Two new, identical cells

Information in each cell could clone you

except for the sperm and egg cells.

Germ cells—sperm for males and eggs for females—also form as you're growing. But they're different, having only half the normal number of chromosomes. Instead of 23 *pairs* of chromosomes, they have just 23, period. Think of these cells as "half cells"; neither can do anything until it unites with the other. Then they combine to produce a cell with a full set of 46 chromosomes, and a new life begins.

The female egg, though tiny, is huge compared to the sperm. Enough sperm to father all people alive today—five billion plus—would fit in a tablespoon with room left over!

At birth, a girl baby's ovaries contain tiny eggs that will be released years later, starting at puberty. All the chromosomes that she'll pass on to her children are in those eggs. A boy will start producing sperm, at puberty, from germ cells present in his testes at birth, and continue to do so throughout his life.

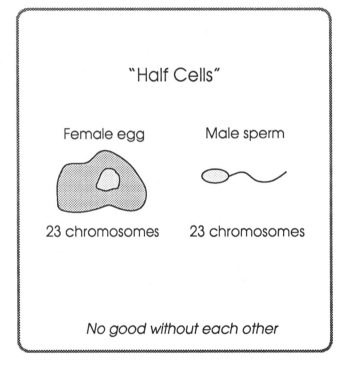

"Half Cells"

Female egg   Male sperm

23 chromosomes   23 chromosomes

No good without each other

Some "coin flips" with your parents' chromosomes . . .

Your father gave you half his chromosomes at your conception; your mother did likewise. This raises a question: Since you didn't get all of your parents' chromosomes—you got only half of them—how were they selected?

When one of your mother's eggs was formed, one chromosome was selected at random (a "coin flip") from each of the 23 pairs in her body cell, to give her egg 23 chromosomes in all. Similarly for your father, giving his sperm 23 chromosomes. When the sperm and egg joined, one chromosome from the sperm and one from the egg combined to give your body cells a pair of chromosomes. This "recombination" occurred repeatedly, forming 23 pairs in all.

So a particular sperm (the lucky one that won the race against millions of others) and a particular egg made you. How many different offspring can two parents have? To make your mother's egg, taking one chromosome from each pair of chromosomes in her body cell yields over eight million possibilities (2^{23}); the same goes for your father's sperm. Combined, there are over 70 trillion possibilities—any one of which could have been you. This is also how many different brothers and sisters you can have.

How is our sex determined? Chromosome pair 23 holds the answer. For this pair, two types of chromosomes are possible, called X and Y. In females, pair 23 has two X chromosomes: in shorthand, XX. Males have an X and a Y: XY. So when the female contributes a chromosome from pair 23, it is always an X. The father's sperm is sometimes an X and sometimes a Y, and this determines the sex of the baby: a sperm with an X produces a girl baby, XX; a Y produces a boy, XY.

Chromosomes are large packages of genes. What are genes? (See pages 14 and 15.) >>>

made you.

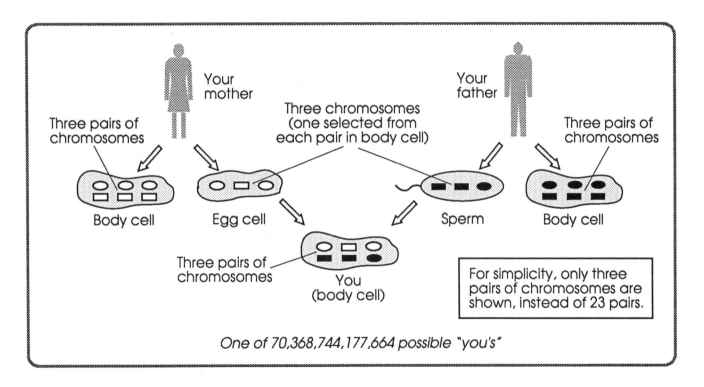

Your mother

Your father

Three pairs of chromosomes

Three chromosomes (one selected from each pair in body cell)

Three pairs of chromosomes

Body cell

Egg cell

Sperm

Body cell

Three pairs of chromosomes

You (body cell)

For simplicity, only three pairs of chromosomes are shown, instead of 23 pairs.

One of 70,368,744,177,664 possible "you's"

A LANDMARK DISCOVERY:

Genes: What are they?

Chromosomes, which are long and narrow, like a piece of thread, are composed of smaller units called *genes*. Now known to be the basic units of heredity, genes work individually or in groups to determine many of our bodily features and even affect our behavior. A series of landmark discoveries beginning in the 1940s found, first, that genes, long suspected of being made of protein, are actually DNA (deoxyribonucleic acid). Then in 1953 Watson and Crick announced that the DNA molecule has the shape of a spiral staircase (double helix). This helped answer the long-standing question of how genes duplicate (see next column). Each gene, of which there are about 100,000 in every cell, is a piece of this spiral molecule.

Peering into chromosomes, we find smaller elements called GENES. These two pages summarize some facts about these elusive items—what they are, how they copy themselves, and what the genetic code is.

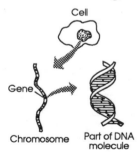

Cell

Gene

Chromosome   Part of DNA molecule

How genes copy themselves:

The Watson-Crick findings helped to explain how genes duplicate. The steps of the DNA "staircase" are nucleic acid bases, labeled A, C, G, T (for adenine, cytosine, guanine, and thymine). The "arms" of the staircase are sugar and phosphate. In DNA, the bases occur in pairs, with A always being joined to T—forming a *base pair*—and C to G. When the cell divides, the DNA molecule splits down the middle, like a zipper. Each half attracts a string of bases similar to the other half, forming two new wholes, each identical to the original molecule.

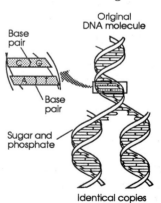

Original DNA molecule

Base pair

Base pair

Sugar and phosphate

Identical copies

THE ELUSIVE GENE IS DNA

The genetic code—how it works:

The order of the bases in DNA determines which protein a gene will build; the protein, in turn, will affect how your body is built. Proteins are composed of smaller units called

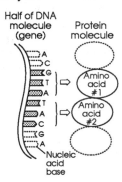

Half of DNA molecule (gene)   Protein molecule

Amino acid #1

Amino acid #2

Nucleic acid base

amino acids. In a *genetic code* common to all forms of life, each group of three bases specifies a particular amino acid: for example, GTA specifies a different amino acid than TAC. Since there are four bases, there are 4 x 4 x 4, or 64, possible triplets, more than enough to designate the 20 different kinds of amino acids. A string of amino acids forms a protein. If the amino acids in one string differ from those in another, the resulting proteins will be different.

A huge amount of information:

To direct the building of your body takes a lot of information. The genes in *each* of your cells consist of about three billion base pairs. The amount of information they hold is enormous. We can make a rough comparison with the English language: In the four-base language of genes, a triplet of bases is roughly equivalent to a letter in English (64 possible triplets vs. only 26 letters, but in English we need upper and lower case, numbers, and some grammatical symbols). So the three billion base pairs, with a billion triplets, correspond roughly to a billion English letters—about the number in 1500 average-sized (300-page) books. Each of the 46 chromosomes in a cell has, on average, the equivalent of more than 20 million letters, or the amount of information in about 33 books.

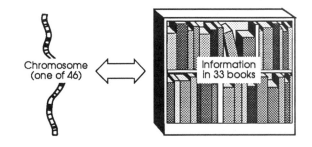

Chromosome (one of 46)   Information in 33 books

Genes work through the genetic code . . .

How does a single cell, by dividing, produce trillions of different kinds of cells—skin, heart, bones, nerves? The food we eat provides energy and raw materials for this feat; the molecules now in our bodies will be gone in several months, replaced by new ones from our food. As the ditty goes:

> It's a very odd thing,
> As odd as can be,
> That whatever Ms. T eats
> Turns into Ms. T.

A cell is a basic building block: it can become just about anything, depending on what the genes, via the genetic code, tell it to do. Although every cell in your body contains all your genes, not all genes are *active* in each cell; many are inactive. The active genes determine what amino acids are selected and what proteins are built: what the cell is—skin cell, liver cell, nerve cell—and ultimately what *you* are.

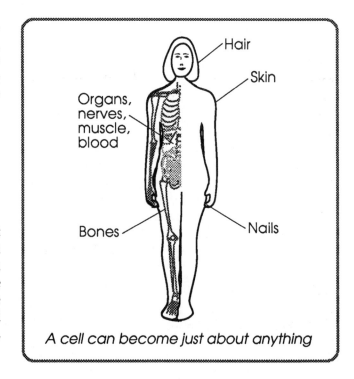

A cell can become just about anything

to make each human being unique.

So your genes spell out how your body is to be built. How do your genes compare to those of other people? For members of your family, we compared groups of genes—chromosomes. But chromosomes tend to get split up with time. For comparison with other people we should look at the genes themselves—or, better yet, at the nucleic acid bases composing the genes. The bases are like a string of beads, which can be strung in different orders. Each of your cells contains about three billion base pairs (the bases occur in pairs in DNA). The number of possible strings is incomprehensibly large (but not all strings would produce a living, healthy human being). You're the only one in the world who has your exact string. The group of proteins made by this string makes you uniquely what you are.

The genes in a single cell, laid out in a line, would stretch about three feet. The genes in all your cells would reach over 400 times from here to the sun!

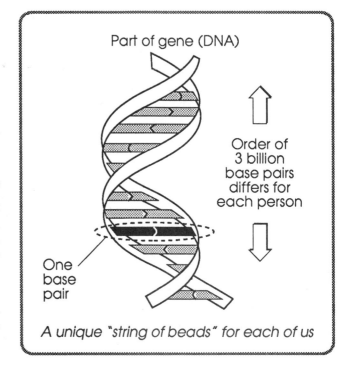

Part of gene (DNA)

Order of 3 billion base pairs differs for each person

One base pair

A unique "string of beads" for each of us

Your genes have been handed down, unchanged,

Your mother, father, and earlier ancestors, like you, also began life as a speck. The speck that was your mother contained the all-important genes from her mother and father. She received half her genes from each of her parents—your maternal grandparents. Your father did likewise, with genes from his mother and father—your paternal grandparents.

Since half your genes came from each parent, and half of *their* genes came from each of their parents—your grandparents—a quarter of your genes, on average, came from each of your four grandparents. (While exactly one-half your genes came from each of your parents, the fraction from each grandparent—one-fourth—is only approximate, because of the chromosome shuffling, or recombination, when germ cells are formed.) You can keep working back: you got about 1/8th of your genes from each of your eight great-grandparents, 1/16th from each of your 16 great-great-grandparents, and so on.

People often have mistaken beliefs about our ancestors. The old ideas about being half Irish, one-fourth German, etc., are a little out of step with the times. Some think their blood is a mixture of their ancestors' blood; but we've seen that even a mother and her baby don't share a single drop of blood. Your blood is made by your genes; it's the *genes* that you've inherited.

The vital genes are sturdy; they're not changed by age, disease, or depression (but they can be damaged by radiation or chemicals). Nor were they produced by your mother and father. Your parents were the temporary custodians of the genes; they just caused some mixing and new combinations when they passed them on to you. Similarly for their parents, their parents' parents, and so on back.

So your genes have been handed down, unchanged, for generations.

for generations.

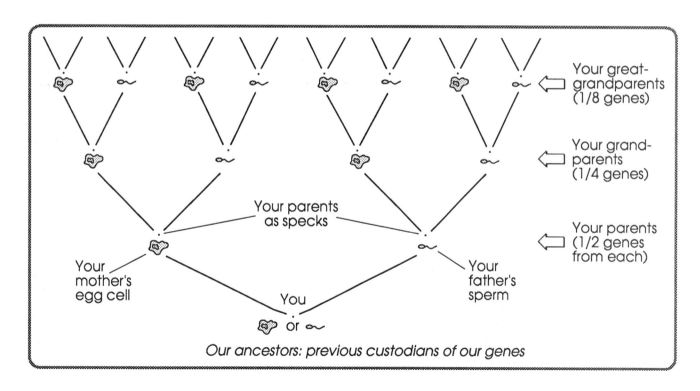

Your great-grandparents (1/8 genes)

Your grand-parents (1/4 genes)

Your parents as specks

Your parents (1/2 genes from each)

Your mother's egg cell

Your father's sperm

You
or

Our ancestors: previous custodians of our genes

Genes, the "heredity" part of the old heredity-vs-environment question, . . .

Which traits are due to heredity (genes) and which to environment? Genes can be traced directly to such bodily features as height, eye color, and skin color, for example. The effects of genes on behavior are not so clear. We see dogs wag their tails, cats purr, spiders spin webs without having been taught. This behavior is instinctive—in their genes. But a cat won't purr if frightened; the environment must cooperate.

The influence of genes for humans is even more tentative, as learning and culture (that is, long-term environmental effects) play a large part. Basic abilities such as intelligence and motor skills are gene-related. But in general, genes seem to confer more of a *potential* than a specific ability. Whether that potential is fulfilled depends—like the cat's purr—on environment.

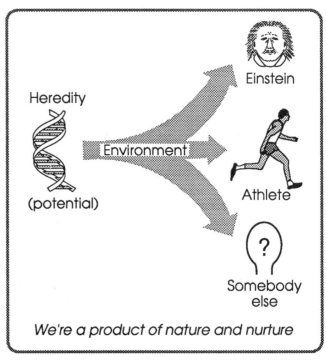

We're a product of nature and nurture

cause differences among races . . .

All humans are members of the same species. They can interbreed, which means their genes are compatible and in this sense are the same for all human beings. There are obvious physical differences among races, in skin, hair, eyes, and so forth. Such differences, due to a relatively few genes, persist for reasons traceable to our past history. Small groups broke away from the main population in the remote past, taking with them somewhat different sets of genes. Over the years, and reinforced by geographical separation, the slight early differences have yielded the racial distinctions of today.

But whatever physical differences persist, they have little to do with mental or artistic capacity. The differences among races in these areas are minor compared to those among *individuals*, regardless of race.

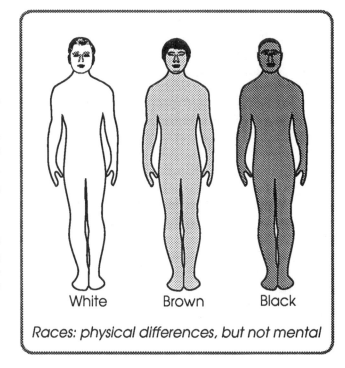

White   Brown   Black

Races: physical differences, but not mental

that are minor compared to environmental effects.

The environment works with what the genes give it; no matter how hard it tries, it can't turn you into a frog. But it can do just about anything else. On TV we see primitive tribes dancing around in weird costumes, chanting gibberish. Strange as it may seem, such "backward" people are our peers, genetically; they have an equal potential for art, literature, science, technology.

Imagine yourself, as a baby, placed with some primitive tribe. You (and any of us) would grow up to fit smoothly into that society (yes, even dancing around and chanting gibberish). Or reverse the scenario: picture a primitive infant raised by a modern family in our society, with our cultural benefits. Our primitive friend would grow up to be just like the rest of us. The plight of primitive people is not brains (heredity), it's culture (environment).

Primitive people—inherently like us

2

FROM FIRST CELL TO ALL LIFE ON EARTH

If uncovering the traces of a distant great-grandparent in a small overseas village fills us with satisfaction, then probing further back to an African ape, a reptile, a fish, that still unknown ancestor of vertebrates, a single–celled forbear, even to the origin of life itself, can be positively awesome.

— Stephen Jay Gould

Our human ancestors stretch back quite a ways, . . .

Looking into the remote past in search of our roots, we find that human beings who look like us and have the same genetic equipment have been on earth for about 40,000 years. Earlier people who were still our same species (Homo sapiens) but who look different go back further, perhaps 200,000 years. Before that were various humanlike or "hominid" species who diverged from us more and more in appearance and brain size the further back we go. Hominids who walked erect, but who had much smaller brains than ours, are known to have lived almost four million years ago. The line of humanlike species split off from the line of apelike (gorilla and chimpanzee) species about five million years ago. Contrary to popular belief, we did not evolve from apes, nor they from us; we did, however, share a common ancestor who lived at that time, some five million years ago.

Where did that common ancestor come from?

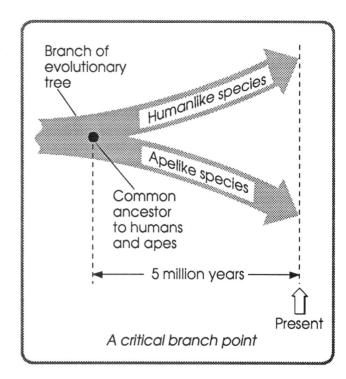

Branch of evolutionary tree

Humanlike species

Apelike species

Common ancestor to humans and apes

←———— 5 million years ————→

⇧ Present

A critical branch point

but it's a short time compared to the age of the earth.

The five million years that humanlike species have been on earth seems like a long time. It's a thousand times longer than the entire period—5000 years—of recorded history. But it's only about 0.1% of the earth's age, which is about 4 1/2 billion years. What happened during the other 99.9% of the time the earth has existed? Was there any life at all? If so, was it related to us?

There was life all right—a great variety. But to trace our roots back further, we need to expand our view of what we are.

As we've seen, people just like ourselves have been here only about 40,000 years. To go back five million years, to the common ancestor of apes and humans, we had to broaden our search to include "humanlike" species. Where did these species and the common ancestor come from? To find out, we need to expand our view still further.

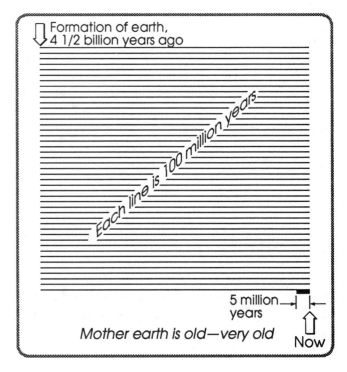

Formation of earth, 4 1/2 billion years ago

Each line is 100 million years

5 million years

Now

Mother earth is old—very old

Life could be traced back "only" about 800 million years, until recently, . . .

We humans (and monkeys and apes) are *primates*, which go back 70 million years. More broadly, we're *mammals*, which first appeared about 200 million years ago. Broader still, we're *vertebrates*, which also include (working backwards) reptiles, amphibians, and fish. Some early *chordate* gave rise to fish 500 million years ago. Continuing on, we're *animals* (not plants or fungi), and finally, we're *living beings*—not inanimate objects. So let's jump to the end of the line: When was the earliest *life* on earth?

Not long ago the earliest life was thought to have been in the "Cambrian explosion," named for the variety of life that erupted 570 million years ago in the Cambrian period. Many fossil shells and bones date from that time. Painstaking search later revealed faint traces of soft-bodied creatures such as jellyfish, sponges and worms, laid down as early as 800 million years ago. Where did they come from? Was there any life before this?

⇩ 4 1/2 billion years ago

?

Traces of sponges, jellyfish, worms (800 million years ago)

Earliest fossil shells and bones (570 million years ago)

Earliest fish   Earliest primates
Earliest mammals

⇧ Now

The dim, mysterious past—was there life?

when discovery of a new type of fossil revised the picture drastically, . . .

In a marvelous decade for biology—the 1950s—a new type of fossil was discovered. The new fossils didn't consist of visible shells and bones but of microscopic life—primeval microbes or bacteria—embedded in rocks. These were much older than 800 million years, the age of the oldest being about 3 1/2 billion years.

With the discovery of this early microscopic life, the date of earliest life on earth was abruptly set back almost three billion years further. The new picture that emerged: for about a billion years (perhaps less), while the earth was cooling, there was no life on earth; then for about 2.7 billion years, only microscopic life; and for the final 800 million years, life that we could see.

Life had taken on a new dimension.

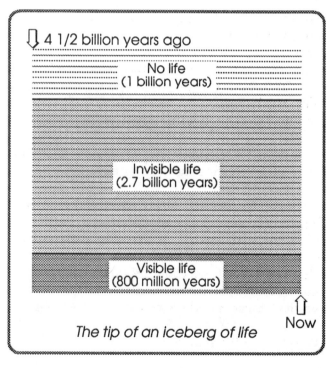

⬇ 4 1/2 billion years ago

No life
(1 billion years)

Invisible life
(2.7 billion years)

Visible life
(800 million years)

⬆
Now

The tip of an iceberg of life

and showed that the only life for almost three billion years was too small to be seen.

The simple fact is that the only life on earth for nearly three billion years was too small to be noticed. It's difficult to get a feeling for such a long stretch of time. Maybe this will help: We're approaching the year 2000. Multiply that by a million, and you have only two billion years. So to reach 2.7 billion, add another 700 million years.

Picture yourself walking around on the earth during that long period: no trees or plants of any sort, no birds, animals—not even insects. A barren, desolate landscape of rocks, lava, sand, water. If you knew what to look for, you'd find a few hints of life: an occasional patch of color or scum on the water or river banks, a faint powder on rocks. You might also have seen strange biscuit–shaped objects near the seashore; these "stromatolites" can be seen today as living fossils in Australia.

What was behind these earliest hints of life?

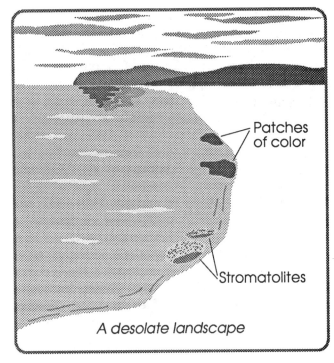

Patches of color

Stromatolites

A desolate landscape

Our earliest ancestor? A pioneer cell . . .

Where did those earliest organisms come from? Many biologists now agree that life arose from a single-celled ancestor at least 3 1/2 billion years ago. (If this startles you, think again of the tiny speck you once were and compare it with what you are now.)

Our earliest ancestor, shown here as a speck, was microscopic, not visible. So this is not an accurate portrait like the one of you when you first started out in life. And there was not necessarily just a single cell. (The "fuzziness" of the boundary between life and non-life is discussed on pages 53-54.) But the actual number is not important; what is important is that this first cell—this *progenote*, as it's been called—represents the conception of not just a single life but of all life on earth. All species, living and extinct, originated here. Like our own beginning, the beginning of life itself was unnoticed and unannounced, and took place in a body of water such as a pond or tidepool (see cover photo) or in the ocean.

Our earliest
ancestor

The conception of all life on earth

that began dividing and kept on dividing.

This first cell began dividing and kept on dividing, just as you did when you were a tiny speck. But for a long time the dividing cells did not cling together as yours did. They remained as separate, microscopic, bacterialike organisms in the primeval seas.

If those early organisms had not survived, we would not be here today to talk about it. Fortunately, old age was not a problem. As single, bacterialike cells, unless they starved or were eaten or injured, they could live indefinitely. Flexible and resourceful, they were able to survive various crises over eons. To this day bacteria still demonstrate the same amazing adaptability: they are found in all kinds of environments, from Arctic ice to boiling hot springs. In recent years they've been used to help clean up oil spills and treat other kinds of waste.

What were these crises that the first cells had to survive? (See next two pages.) >>>

Pioneer cell

The pioneer cell multiplied by dividing

THE FIRST CELLS: MEETING

The first problem was food:

To survive, life must have food. The sea initially had a rich store of organic (complex carbon-containing) molecules, from which the cells got energy by fermentation—a process still used by bacteria (giving us cheese, vinegar, yogurt, and other products). Other cells were able to reduce their need for organic molecules, which were disappearing rapidly, by generating energy from inorganic molecules. Today such "methane bacteria" produce methane (or marsh gas) from sewage and sludge.

The first cells faced the problem of surviving in an ocean that had never before supported life. For more than two billion years, they survived a series of crises—if they hadn't, we wouldn't be here today. These two pages summarize their story.

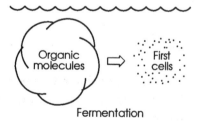

Fermentation

When food ran short— photosynthesis:

Food supplies began to dwindle, threatening life's tentative beginning with famine. Some enterprising bacteria created their own food from sunlight, hydrogen, and carbon dioxide—they "discovered" photosynthesis. Hydrogen came first from hydrogen sulfide; later, it was taken from the largest source available, water. This feat of photosynthesis—with bacteria creating their own food—must be emphasized. *It's considered by many biologists to have been the most important step in the evolution of life.* (Plant photosynthesis came much later.)

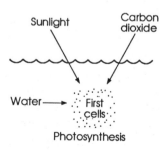

Photosynthesis

A SERIES OF CRISES

Oxygen in the air led to respiration:

Removing hydrogen from water (by photosynthesis) released oxygen, which began to accumulate in 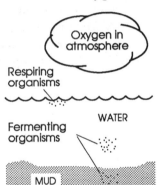 the atmosphere. Today, in ancient Arizona rocks you can see oxidized iron bands formed 2.5 billion years ago from oxygen released by bacteria. In those times, unlike today, oxygen was poisonous to all living cells. To escape, bacteria had to take refuge in deeper water or mud. Or they had to learn to live with oxygen. Again, some bacteria met the challenge; they not only learned to live with it, they used it to advantage. They found a far better way than fermentation of breaking down organic molecules—respiration.

Another milestone—nucleated cells:

Bacteria cells are quite primitive compared to all multicelled life. Whereas in our cells the genes are packaged in chromosomes contained in a nucleus, bacterial cells have a single strand of DNA and no nucleus. These simple cells had been around for one to two billion years when a new type of cell appeared, which was larger, more complex, and, most importantly, had a nucleus. The distinction between these two types of cells—one with a nucleus and the other without—is critical. It is, biologists say, *even more fundamental than the distinction between animals and plants.* The reason is 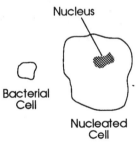 clear: Whereas bacteria have not changed for eons, nucleated cells were the harbingers of familiar plants and animals, including us.

The start of something big: a remarkable symbiosis . . .

Life was still invisible about 1 1/2 billion years ago, but two types of cell now lived side by side: bacteria-like cells with no nucleus, and new, larger, nucleated cells. Then there was a remarkable symbiosis in which the bacteria cells invaded and became a permanent part of the new cells. Such associations are not unusual; for example, a single-celled organism lives inside the termite, and helps it digest wood.

With the aid of their new bacterial partners, the nucleated cells could now breathe oxygen—they found respiration. Some acquired, in addition, photosynthesis. The first group were the forerunners of animals; the second, of plants. The remnants of these ancient bacteria are still in modern animals and plants and are known as mitochondria and chloroplasts, respectively. They have their own genes, as discovered in the 1960s. Every cell of our bodies harbors these mitochondria—replicas of bacteria of billions of years ago.

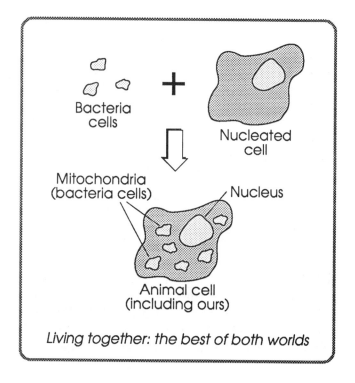

Bacteria cells

Nucleated cell

Mitochondria (bacteria cells)

Nucleus

Animal cell (including ours)

Living together: the best of both worlds

that led eventually to multicelled life.

The new nucleated cells and respiration formed a powerful combination. Nucleated cells were the kind of building blocks needed for larger bodies, and respiration was much more efficient than fermentation in generating the necessary energy. Together they spurred on a new form of life, multicelled organisms. Such organisms enjoy many advantages over single-celled types. Being larger, they're less in danger of attack by other organisms. Most important was that, with many cells available, some could be used for special purposes such as nerves or a digestive system. Age, however, now entered the picture: the endless lifespan of single cells was not conferred on larger organisms.

Until now, life had been invisible. The appearance of jellyfish, sponges and worms announced the arrival of visible life in the sea. They were followed by shellfish and other creatures with hard body parts that left a clear record for us to examine.

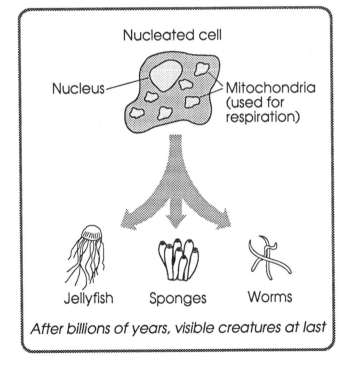

After billions of years, visible creatures at last

Protective ozone allowed life to move to land, . . .

Early life had been confined to the sea. The land, barren and forbidding, was exposed to the sun's deadly ultraviolet rays. In time, as oxygen accumulated from bacterial photosynthesis, a layer of ozone formed in the upper atmosphere, shielding the land from the lethal rays. Life—in turn plants, lobe-fin fish, and amphibians—seized this opportunity to begin to explore the land.

Compared to the sea, the land proved to be a harsh environment indeed. Animals and plants had to find ways to preserve liquid in their bodies to avoid dying of thirst. Eggs needed more protection. Cold and heat extremes, unknown in the sea, posed a new threat. Gravity, which had been all but unnoticed in water, made its presence felt and required strong bones for bodily support.

Life in the vanguard had to adapt to these new conditions or perish.

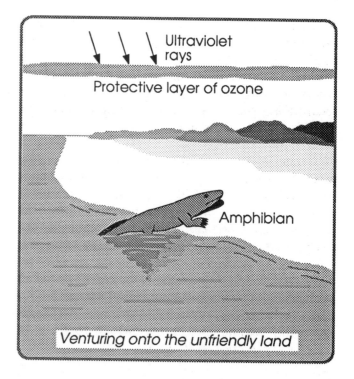

Ultraviolet rays

Protective layer of ozone

Amphibian

Venturing onto the unfriendly land

where a variety of life evolved . . .

When animals and plants were able for the first time to leave the sea and reach out to land, different varieties evolved at a more rapid pace. Plant seeds and spores probably led the way, some 450 million years ago. An ever-expanding tree of life followed, bearing a great variety of fruit—amphibians, reptiles, birds, and mammals (including us).

Why such a variety? Wherever there's a "niche"— a unique set of local environmental conditions— there will be forms of life competing for it. Over vast stretches of time, gene mutations (copying errors) led to diverse life forms to exploit the many niches that the world offers. Sexual recombination of genes accelerated the evolution of an expanding variety. Those forms that were most adaptable to prevailing conditions and capable of reproducing survived through natural selection; others succumbed. In time we're all vulnerable; 99% of all species that have ever lived have become extinct.

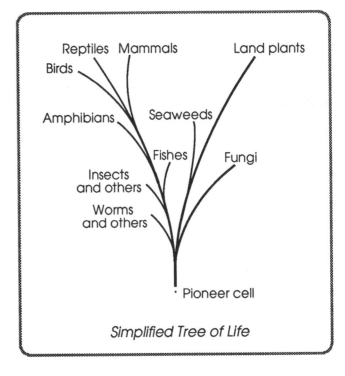

Simplified Tree of Life

into many forms and species, including us.

Life on earth today consists of millions of species. (Put simply, a species is a group whose members interbreed and produce fertile offspring.) By far the greatest number of species are insects, with over 20,000 butterflies alone. But even within a single species there is variety; members of a species don't always look alike. You can see this clearly with domesticated animals such as dogs and cats, which have been selected over the years by humans for certain distinguishing features (artificial selection, as opposed to natural selection). From St. Bernard to Chihuahua, dogs are one species. Similarly for domesticated cats: Persian, Siamese and others are all the same species. Among wild animals, Asian and African elephants look similar (the African has larger ears) but are two distinct species.

Humans, whether Pygmies or Eskimos, white, black, red, yellow or brown—plenty of variety, certainly—are all members of the same species.

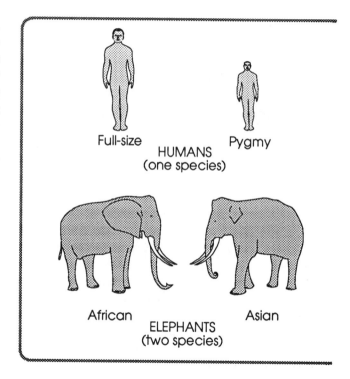

Full-size HUMANS Pygmy
(one species)

African ELEPHANTS Asian
(two species)

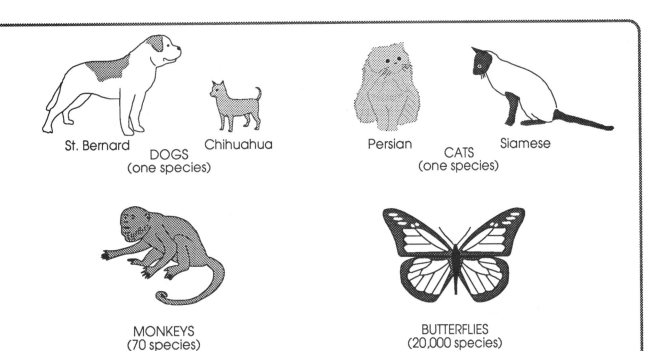

St. Bernard   DOGS
(one species)   Chihuahua

Persian   CATS
(one species)   Siamese

MONKEYS
(70 species)

BUTTERFLIES
(20,000 species)

3

THE ORIGIN OF LIFE

The past 30 years have seen tremendous progress in the ability to conjecture about the origin of life on earth.

— James D. Watson

How did life originate? We need to focus on the first cell, . . .

Sooner or later, in tracing back our roots, we must face the awesome question of how life originated. This question may seem impossibly difficult if we look at life around us today and ask where it all came from. But let's remember that we've already covered a lot of ground. We've followed life all the way back to a microscopic speck—our earliest ancestor, the ancestor of all life.

In working our way back, we found that the visible, more familiar forms of life such as plants and animals didn't just spring into existence. They were preceded by billions of years of invisible life such as bacteria. And what started it all—our earliest ancestor, the pioneer cell—was also invisible, microscopic. So in thinking about the origin of life, we need to focus on that original speck, that first cell.

How did it arise?

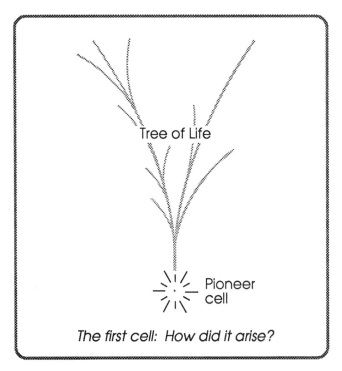

Tree of Life

Pioneer cell

The first cell: How did it arise?

which was too complex to have formed spontaneously, . . .

Even the "simple" bacterialike cell—similar to the earliest cells on earth—is quite complex, much too elaborate to have arisen spontaneously. But complex as it is, there is no magic. The internal activities are based on the same laws as those that describe molecular behavior outside of cells.

What keeps the cell going? To oversimplify what is an intricate series of operations: Reactions requiring energy, such as those building large molecules, are coupled with reactions that release energy, with the entire system tending to seek stability. Cell activities are like other physical events—for example, water running downhill—in seeking such stability, or "state of minimal energy." The spherical shape of our cell membranes illustrates the principle, as do the unique folding patterns of protein molecules.

If not spontaneously, how *did* such a complex object as a cell originate?

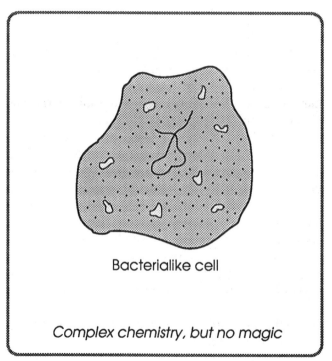

Bacterialike cell

Complex chemistry, but no magic

and must have been preceded by simple building blocks.

To form such a complex thing as a cell, even the simplest cell, there must first have been an accumulation of simpler cell "building blocks." What kind? At the first level of breakdown, cells are composed mainly of nucleic acids (DNA and RNA—ribonucleic acid), proteins, and fatty material for the enclosing membrane. But nucleic acids and proteins are themselves large, complicated molecules. So let's break them down further: nucleic acids are made up of bases, phosphates, and sugars, while proteins consist of chains of amino acids.

The mere existence of such building blocks—smaller molecules—does not guarantee that cells will ever be built. But it's a logical first step to see if even these more elementary materials could originate naturally on the primitive earth.

So the next question we must ask is: What was the primitive earth like?

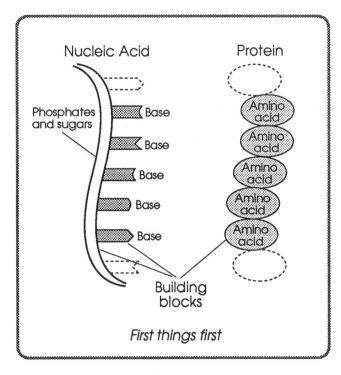

First things first

Experiments simulating early-earth conditions . . .

The primitive earth was completely different from what it is today. All the water in the oceans was in the atmosphere as steam; as it condensed, it rained down, only to be recycled aloft as it reached the hot surface. The turbulence would have caused vast amounts of static electricity...and lightning. No ozone layer protected the earth's surface from the sun's energetic ultraviolet rays, as the atmosphere was without oxygen. The earth was probably a cauldron of activity as the shallow, warm seas were bombarded with lightning, ultraviolet rays, volcanism, and radiation from the earth itself.

Various experiments have tried to reproduce such early-earth conditions in the laboratory. Different mixtures of gases have been used since, though it's known there was no oxygen, the exact composition of the earth's atmosphere at the time is uncertain.

The first such experiment has since become famous.

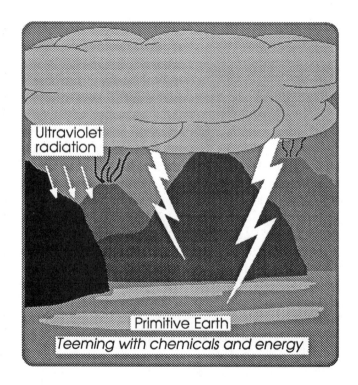

Ultraviolet radiation

Primitive Earth
Teeming with chemicals and energy

have produced cell building blocks in the laboratory.

The Miller-Urey experiment of 1953 was a milestone in showing what could be produced under reasonable, simple conditions in the laboratory. A mixture of gases (methane, ammonia, hydrogen, water vapor) thought to be common in the early-earth's atmosphere was subjected to an electric spark, representing an energy source such as lightning. Much to everyone's surprise and delight, amino acids and other organic compounds were produced.

Later experiments confirmed these findings, and even were successful with different mixtures of gases, provided carbon, hydrogen, oxygen, and nitrogen were present. (The oxygen must be in a combined form such as water vapor or carbon dioxide.) These four elements, which form 98% of our bodies, are (along with helium, which is found in stars) the most common elements in the universe.

A good first step. But important questions remain.

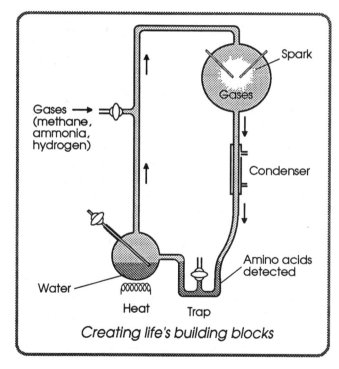

Creating life's building blocks

But big questions remain: How are building blocks ordered . . .

The creation of building blocks in the laboratory, while an exciting first step, still leaves us a long way from a living cell—sort of like a truckload of bricks compared to a finished building. The cell building blocks must be linked together to make the larger molecules (nucleic acids and proteins) and the exact order of the linkages is crucial. In today's cells, the protein building blocks and their order—which together are what distinguishes one protein from another—are specified by the nucleic acids. In a complicated sequence of steps, the information in DNA is transferred by several types of RNA to where it is needed in the cell to direct the job. The order of nucleic acid building blocks is determined by copying what's already there, which for the most part was arrived at long ago by natural selection. The copying is done with the help of proteins.

If it's beginning to sound as if we're going around in a circle, you're getting the picture.

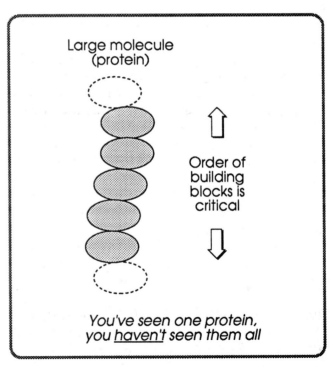

Large molecule
(protein)

Order of
building
blocks is
critical

*You've seen one protein,
you <u>haven't</u> seen them all*

and how are they joined?

With their linkage order dictated by nucleic acids, the building blocks must still be joined together to form the larger molecules—and they don't readily do this. In existing cells the joining is done with the help of enzymes. An enzyme, like most of the stuff of our bodies, is a protein. Enzymes are truly versatile molecules: "An enzyme can do just about anything," as one biologist puts it. It works its wonders by means of the shape that it folds into, which is determined by its sequence of amino acids. The shape creates a binding site that can bring two molecules together in such a way that they will react, whereas normally they would not (or would do so very slowly). In this way all larger molecules are put together—DNA, RNA, even enzymes themselves.

With nucleic acids and proteins so mutually dependent, how could things even get started? Which came first? (See next two pages.) >>>

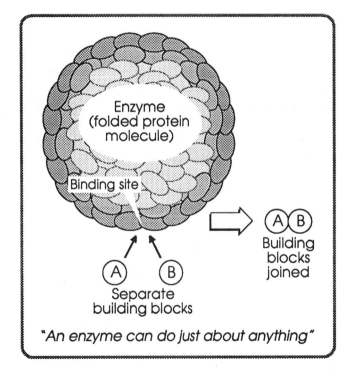

Enzyme
(folded protein molecule)

Binding site

A B
Building blocks joined

A B
Separate building blocks

"An enzyme can do just about anything"

A CHICKEN AND EGG RIDDLE:

Maybe neither nucleic acids nor proteins came first:
They may have evolved together. Like cells, these large molecules are too complex to have just appeared spontaneously. They may have been preceded by a gradual evolution of short strings of amino acids and nucleotides. Such primitive strings might have helped each other through "autocatalysis": a primitive nucleic acid directing the building of a protein which in turn helped build the nucleic acid itself. When nucleic acid or a primitive predecessor was able to reproduce itself—*replicate*—like this, life was not far off.

A hard nut to crack in the origin of life is which came first, nucleic acids or proteins? Nucleic acids (DNA and RNA) direct the building both of themselves and proteins, while proteins, as enzymes, help DNA and RNA do their job. These two pages outline the speculations.

Building directions

Nucleic Acids ⟷ Proteins

Building help (catalysis)

Which came first?

DNA probably was a late arrival:
Simple proteins and RNA probably preceded DNA. For one thing, the components of proteins (amino acids) form more readily than the components of nucleic acids (nucleotides). Similarly for the ribose sugar of RNA vis-a-vis the deoxyribose of DNA. Also, the uracil of RNA is much more easily damaged than its counterpart in DNA (thymine), leading to a much higher mutation rate for RNA. Another clue: certain types of RNA can, like enzymes, work as catalysts—something DNA can't do. So an archaic RNA could have both copied itself, like DNA, and catalyzed other reactions, like protein.

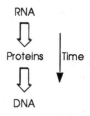

RNA

⇩

Proteins | Time

⇩

DNA

Probable order of appearance

WHICH CAME FIRST?

Clay crystals—primitive genes?

An even simpler mechanism may have come earlier. Perhaps clay crystals. Made of silica, the commonest substance on the earth's surface, they have the following characteristics: They reproduce, an essential of life. They sometimes have flaws, which also reproduce; such "mutations" are necessary for evolution. Finally, they can serve as catalysts by concentrating chemicals, helping them to react. Experiments show that microscopic clay crystals attract amino acids; and the amino acids are, like those in living things, all of the same stereoisomer, or "handedness." (Amino acids have been found in meteorites, but they have been of mixed isomer.) Such crystals may have served as primitive genes, evolving later into the more elaborate RNA/DNA.

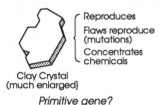

Reproduces
Flaws reproduce (mutations)
Concentrates chemicals

Clay Crystal (much enlarged)

Primitive gene?

A vital necessity—protective skin:

Regardless of how an evolving cell was formed, it would need a protective "skin" to concentrate chemicals and allow reactions to continue and build on each other. When a mixture of fat and water is shaken, small, spherical particles form, which have a double-layered skin or membrane. The fatty skin formed in this simple way is the same kind that protects cells for all living things today. Some argue that such pseudo-cells, or "coacervate" droplets, arrived before anything else. They allowed the concentration of elements necessary to form the more complex molecules such as nucleic acids and proteins.

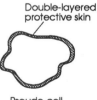

Double-layered protective skin

Pseudo-cell

All three major elements—skin, nucleic acids, and proteins (enzymes)—have their adherents as to which came first.

Another hurdle: A living cell is more than nucleic acid and proteins.

Even if nucleic acids and proteins were present, how did they evolve into the much more complex, living cell? Experiments show "evolve" is the right word here; a type of evolution almost certainly occurred prior to any life. What is necessary for evolution is not life but variation, selection, and reproduction.

In one such experiment, nucleic acid and proteins (enzymes)—completely inanimate materials—when placed in a test tube were found to reproduce like bacteria cells. A poison was then added to see how the mixture would respond. Reproduction stopped, but then continued after a while. A nucleic-acid mutation had made survival possible, just as some bacteria survive when exposed to antibiotics.

This clever experiment showed that once a repro-ducing system of nucleic acids and protein formed, it could have evolved into something behaving like a living cell.

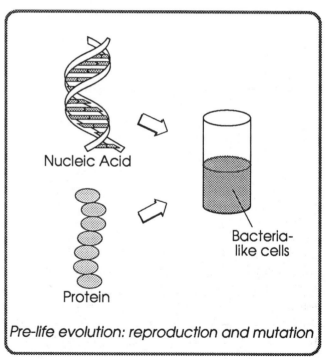

Nucleic Acid

Protein

Bacteria-like cells

Pre-life evolution: reproduction and mutation

When did life begin? There's no clear boundary, . . .

However achieved, a living cell probably arrived only after a gradual process punctuated by many failures. In an environment full of energy from various sources, there was ample time and opportunity for many combinations of chemicals to form, break up and reform in different ways. Eventually the autocatalytic type of reaction, where one type of molecule helps create another that in turn assists the creation of the first, worked together with evolutionary forces to lead the way to life.

When did life actually begin? It's unlikely that there was ever a distinct moment when this occurred, a clear boundary between life and non-life. As Nobel laureate John Kendrew has said: "I do not think there is any evidence of such a boundary, of any difference between the living and non-living, and I think most molecular biologists would share this view."

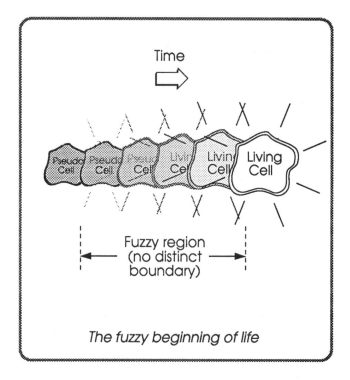

The fuzzy beginning of life

as today's viruses bear witness.

Bacteria, though small and simple, are unquestionably *alive*. The much smaller and simpler viruses, however, pose a question: Alive or not? Inside our cells, they seem only too alive, reproducing and causing diseases such as measles and smallpox. But they can also exist as lifeless crystals.

A virus is a molecule of RNA or DNA with a protective coat. Some very simple types, called viroids, are just RNA without even a protective coat; they may represent life at its simplest. Besides causing disease, viruses are suspected of carrying genes between species—"jumping genes." Some biologists raise the intriguing possibility that viruses may just be ancient pieces of genes from other forms of life.

Whatever they are, viruses illustrate a kind of problem we keep running into: life vs. non-life, human vs. non-human, embryo vs. human being. There just are no clear, sharp boundaries.

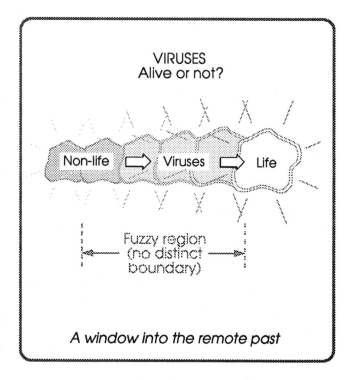

A window into the remote past

For life to appear on earth, there had to <u>be</u> an earth, . . .

Earlier we spoke of the primitive earth. How did it originate?

The solar system, including the earth, formed about 4 1/2 billion years ago from a mixture of gases containing hydrogen, helium, nitrogen, carbon, oxygen, and other elements. The earth was inhospitable to life until the surface had cooled to a solid form and water had rained from the atmosphere to form the oceans. It's now believed that the earliest the earth could have supported life was about four billion years ago. Microscopic fossils have been dated to about 3 1/2 billion years ago, and there is some indication of even earlier life, at 3.8 billion years. It therefore seems that once life was possible, it appeared after "only" 200 – 500 million years—not a long time, geologically speaking.

We're not through yet. Where did the original gases that formed the earth come from?

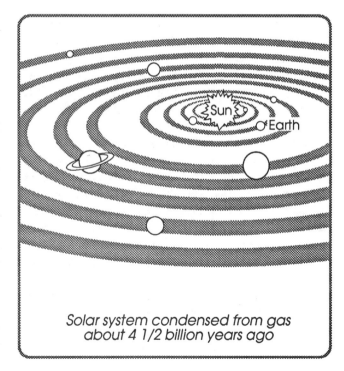

Solar system condensed from gas about 4 1/2 billion years ago

which formed from elements originating in supernovas, . . .

The elements that formed the solar system and earth originated from cataclysmic supernovas—exploding stars—billions of years earlier. Though rare, supernovas are still observed today. The force of gravity condenses huge clouds of gas; thermonuclear reactions in the hot center of the condensing mass create the heavier elements from the lighter hydrogen and helium. The condensation reaches enormous densities, first forming a star and then exploding with unbelievable force. A relatively close supernova was observed in 1987 in the Large Magellanic Cloud, 160,000 light-years away; in the first ten seconds it released 100 times as much energy as the sun will emit in its entire ten-billion-year lifetime! The heavier elements scattered by such explosions eventually form other stars and planetary systems like our own.

We're almost to the end of the line. What created the original stars and supernovas?

Where our atoms came from

which arose from hydrogen and helium created in the Big Bang.

The final step that we can take backward is to the Big Bang, the origin of the universe. The universe is expanding; stars and galaxies are moving away from each other at rapid speeds. These movements can be projected back to a time some 15 – 20 billion years ago when, in the words of the noted physicist Stephen Hawking, "the universe was infinitesimally small and infinitely dense." It was a pinpoint of energy, a "singularity" where all physical laws and predictability broke down and time itself began.

Some time after the Big Bang, single protons and electrons came together to form atoms of hydrogen, the commonest and simplest element in the universe. Helium, the next simplest and commonest element, followed, and the universe as we know it was on its way.

What happened before the Big Bang? We'll leave that as an exercise for the reader.

The universe just after the Big Bang

The beginning of time

4

OUR KINSHIP WITH OTHER CREATURES

Such an astonishing degree of uniformity [of all living things] was hardly suspected as little as 40 years ago.

— Francis Crick

Our early life: A story within a story.

OUR REMARKABLE GROWTH from speck to baby harbors another surprising tale—a sort of epic *déjà vu*—that sheds light on our roots in the dim past. Beginning as a speck that corresponds to the first cell on earth, we pass through stages in a matter of months that retrace those that life itself experienced over billions of years. Though the correspondence is far from exact, it's close enough to have intrigued scientists over the years, starting with Ernst Haeckel in the last century and continuing more recently with the noted biologist Stephen Jay Gould.

Examining our prenatal growth, we're struck by similarities with earlier life forms. Our curiosity aroused, we probe deeply into our cells—our "inner space" (Amram Scheinfeld). There we find ourselves looking back in time, just as we do when we gaze into outer space and see stars not as they are now but as they were thousands, millions, billions of years ago. But we're getting ahead of our story.

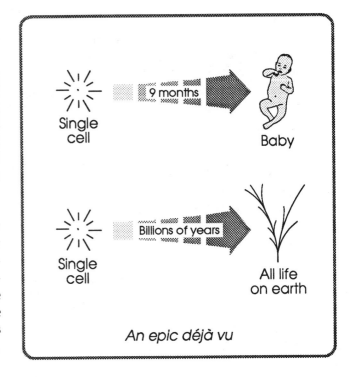

Single cell — 9 months → Baby

Single cell — Billions of years → All life on earth

An epic déjà vu

The stamp of evolution is on our early growth, . . .

Let's go back: the speck you once were starts to divide. Like your earliest ancestors, the first cells look the same and are independent; if two of them separate, they become identical twins. At this stage they act as a colony of cells, not a multicellular organism (this also explains why each of our cells has all the genes). Tissues form, as the growing ball of cells, like an ancient invertebrate, is enveloped in a sac of fluid similar to the primeval sea. Rudimentary, fishlike gill slits appear. Organs take shape: A primitive, fishlike kidney forms, to be replaced by a second and then a third less-primitive type. A growing tail reaches its maximum length during the second month, then disappears. A notochord resembling that of an early ancestor is eventually replaced by a jointed backbone. Your eyes, originally on the side of your head like those of a fish or reptile, later move around to the front. At birth, like the earliest amphibian crawling out of the sea, you leave your watery home and enter the outside world.

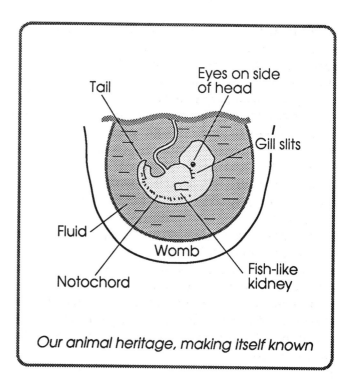

Our animal heritage, making itself known

and is evident elsewhere, . . .

In addition to our own prenatal life, many other signs point to a common origin for all life. A human cell resembles the same type of cell—blood, heart, muscle, nerve—in other animals more than it does a different type of human cell. Our bone structure resembles that of many other animals, from birds to whales (more on this shortly). Our origin, and that of all animals, in the ancient sea is reflected in our body's requirements for moisture. Our body fluids meet the needs of body cells and germ cells for a watery environment. Body cells are bathed in a proper mixture of water and salts. Germ cells also must have moisture, whether in spawning or internal fertilization. We even need moisture to breathe. We can't use oxygen directly; our lungs need a surface layer of moisture to dissolve it first. So do most amphibians: beginning life as tadpoles, which first breathe with gills, they then move to land and breathe with lungs as did their ancestors of millions of years ago.

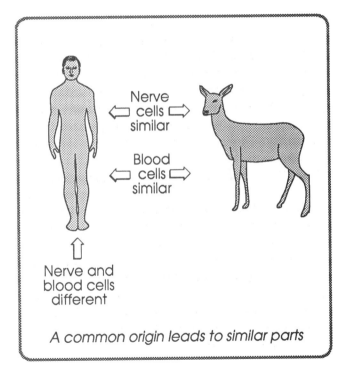

Nerve cells similar

Blood cells similar

Nerve and blood cells different

A common origin leads to similar parts

even inside our cells.

The unity of life continues right on down to the molecular level. For all varieties of life, nucleic acids are composed of the same four nucleotides, proteins of the same 20 amino acids, and the same code translates from one to the other. Compatible segments of DNA have been exchanged between such diverse life forms as mammals and bacteria. All proteins are made in tiny ribosomes. All cells have the same molecular symmetry (left-hand for amino acids and right-hand for nucleic-acid sugars) and the same ancient fermentation process, with energy stored in a universal molecule, ATP (adenosine triphosphate). The membrane enclosing cells is universal to all life, ancient and modern. The same mitochondria of primeval origin appear in the cells of animals, plants, and fungi, where they oxidize food for energy.

The cell is a warehouse of relics from the remote past, all pointing to life's common origin.

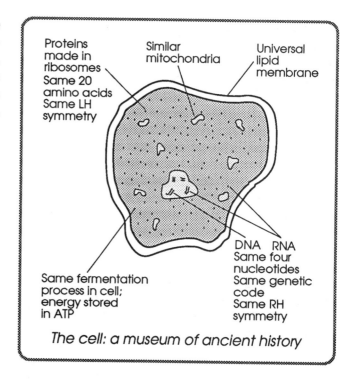

Proteins made in ribosomes
Same 20 amino acids
Same LH symmetry

Similar mitochondria

Universal lipid membrane

Same fermentation process in cell; energy stored in ATP

DNA RNA
Same four nucleotides
Same genetic code
Same RH symmetry

The cell: a museum of ancient history

Early life on earth could become just about anything . . . and did.

Life had its common origin in a single cell, which continually divided. For almost three billion years, as we've seen, life was restricted to single, microscopic cells. Not much variety there. How could the kind of diversity that we see around us now—animals, plants, fungi—possibly have arisen from such modest beginnings?

Let's recall the vast difference between what we look like now and the original speck or seed we sprouted from—our own modest beginnings. We look inside our cells now and find the same genes that were in that speck. Similarly, the early cells that started life have persisted as an enduring thread, providing an underlying unity to the growing evolutionary tree. The great variety in the branches and twigs is a product of an evolution that has been highly opportunistic, seizing the chance for invading any favorable niche that would support life.

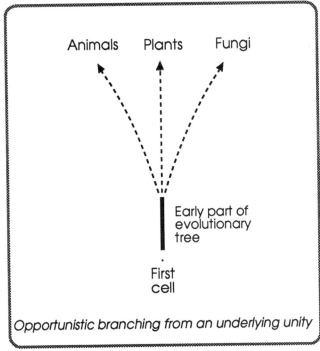

Opportunistic branching from an underlying unity

Similarly, when we're very young we all look alike; . . .

At the beginning, like the first cells of the evolutionary tree, we all look the same: All life, from single-celled amoeba to 100-ton whale, with us in between, begins life as a tiny speck. No variety. Later, while in the early-embryo stage shown here, we still have more than a passing resemblance to other creatures. Even experts have trouble telling one from another.

older, we look different.

As we get older, like twigs of the evolutionary tree, we look quite different. Even a baby is a "little human being," with a human face, arms, legs, . . . Puppies and kittens look like little dogs and cats. What was once indistinguishable is now, fortunately for all of us, easily identified even by non-experts.

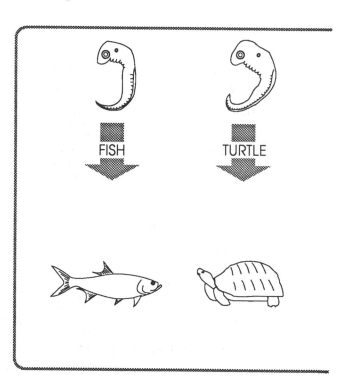

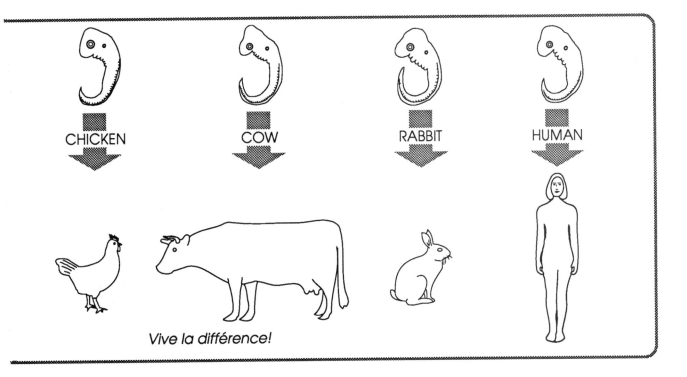

CHICKEN   COW   RABBIT   HUMAN

Vive la différence!

Evolution has been a great "tinkerer," . . .

Evolution has been highly opportunistic in exploiting different niches for life. But underlying this trend has been a basic conservatism. Once changes are made, they tend to be retained; there's no going backward. Evolution builds on what has gone before, reinforcing the unity of life stemming from a common origin. Here are two examples:

- **Fermentation/respiration**: When respiration was discovered by early cells, it didn't replace the older fermentation but was simply added on. Our bodies still retain the primitive fermentation ability in addition to normal respiration.

- **The panda's thumb**: The panda descended from meat-eaters, but now eats only bamboo. A "thumb" evolved from the panda's wristbone, making it easy to hold bamboo shoots.

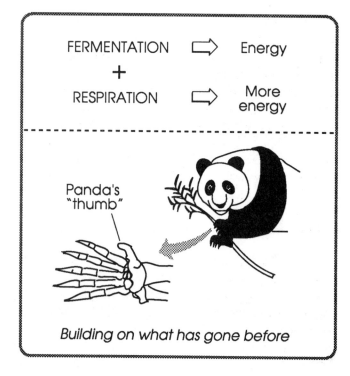

Building on what has gone before

making do with what's available . . .

The bones of our arm fall into three groups: a top section (humerus), a middle section (radius and ulna), and a bottom section of wrist and hand. These same bones, in roughly the same relative positions and even with the same names, are also found in a bird, dolphin, dog, whale, and other animals. Even reptiles—those with limbs, such as turtles and alligators—have the same three-part structure. Though the pattern is similar, the bones are not the same in size and appearance for all these animals. And the number of bones in the "wrist" and "fingers" may differ; for example, birds have only two wrist bones, while most mammals have seven or eight.

The bone arrangements arose from a common ancestor but have been modified to perform different tasks: to fly, to paddle, to grasp. The common ancestor, an early land vertebrate, had four limbs similar to those shown here, with the same bones arranged in the same way.

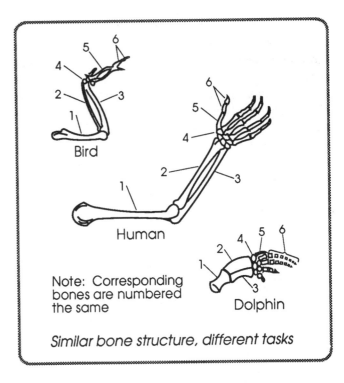

Bird

Human

Dolphin

Note: Corresponding bones are numbered the same

Similar bone structure, different tasks

in meeting life's challenges.

That evolution is conservative and builds on what exists rather than invent something new is shown by many examples:

1. **Mammals' jaws:** Mammals evolved from reptiles, which have five bones in the lower jaw; the jaw of mammals has only one. The missing bones are part of our middle ear—a change made over time to improve the hearing of reptiles. So our jaw, which is much better for chewing than the reptile jaw, turns out to be a side product of the evolution of better hearing!

2. **Arms and legs:** Why do we have two arms and legs? Because early fish had two pairs of fins, which became rudimentary legs when fish evolved into amphibians to invade the land. The legs of amphibians were eventually passed on to four-footed mammals and to us.

3. **Bird feathers:** Birds evolved from small feathered dinosaurs (so the tiny canary and multi-ton brontosaurus are more closely related than you might think). These dinosaurs used feathers as insulation for body temperature control, or to scoop up small prey. Feathers were eventually put to good use by birds for an entirely different purpose—flight.

4. **Orchid pollination:** Orchids adapted standard parts of ancestor flowers to build special devices to insure cross-fertilization. These creations attract insects, cause sticky pollen to adhere, and insure that the pollen comes into contact with female parts of the next orchid the insect visits.

Evolution is an intriguing mixture of opportunism and conservatism. What is the driving force behind it? (See pages 72-73.) >>>

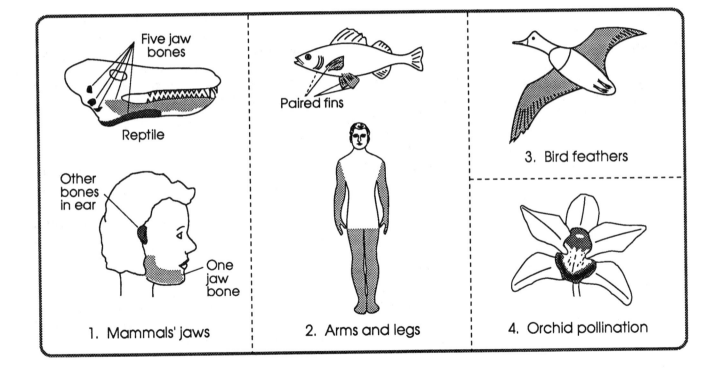

Five jaw bones

Reptile

Other bones in ear

One jaw bone

1. Mammals' jaws

Paired fins

2. Arms and legs

3. Bird feathers

4. Orchid pollination

MUTATIONS: COPYING ERRORS

What is a mutation?

Cells divide and their genes are copied, again and again. On rare occasions the copying is incorrect, producing a *mutation*. The mutated gene then reproduces itself in its new form. Mutations occur at random and are usually harmful, producing such diseases as sickle-cell anemia and cystic fibrosis. Since they introduce variation that may be useful in a changed environ-

If the first cell had kept on reproducing accurately, there never would have been anything but single-celled life. Mutations introduce changes which, along with recombinations (pages 12, 37), drive evolution. Mutations are explained on these two pages.

ment, however, some mutations can lead to new, successful forms of life. We all carry mutant genes, but most are "recessive" and are harmless because of the good gene in the other chromosome of the pair.

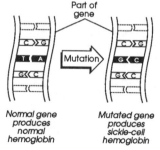

Part of gene

Mutation

Normal gene produces normal hemoglobin

Mutated gene produces sickle-cell hemoglobin

Mutations can be traced through time:

In various altered forms over eons, our genes have appeared in jellyfish, worms, fish, reptiles, early mammals, monkeys and apes. So we share genes with other creatures, the closer the relation, the more shared: 58% with lemurs, and an amazing 97.5% with the chimpanzee. In tracing mutations, one human protein was found to differ in four amino acids

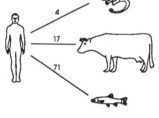

(out of 150) from that of a monkey, 17 from a cow, and 71 from a fish. Humans separated from monkeys 25 million years ago, from cows 90 million years ago, and from fish 400 million years ago.

THAT HELP DRIVE EVOLUTION

A new discovery—regulatory mutations:

A mutation typically occurs at a single point of the chromosome—a single nucleic acid base. It affects the building of a protein, and results in a structural defect in the body. In addition to such "structural" mutations, another type— "regulatory" mutations— have recently been discovered. As their name implies, these affect how and when the proteins are put together. Usually more than just a point is affected; perhaps a large segment of a chromosome is inverted, for example.

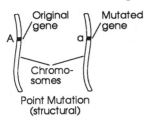

Original gene Mutated gene

A a

Chromosomes

Point Mutation (structural)

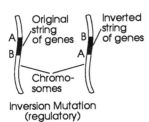

Original string of genes Inverted string of genes

A B

B A

Chromosomes

Inversion Mutation (regulatory)

These mutations are thought to have had a strong influence in shaping evolution (see next column).

Mutations are critical in early life:

Animals are quite malleable when young ("as the twig is bent, so grows the tree") and become more specialized and less flexible as they mature. Evolution, ever opportunistic, has used regulatory mutations to slow or speed up the rate of maturity to produce different life forms to exploit available niches. For example, a salamander, an amphibian, normally leaves water for land on maturity. If the pond is attractive, it doesn't "want" to leave to face the harsh land. A form called an axolotl has evolved that, though sexually mature, remains in a permanent juvenile state, staying in the water and retaining its larval gills.

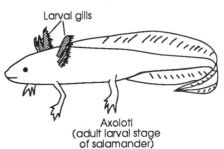

Larval gills

Axolotl (adult larval stage of salamander)

Mutations help to explain our own ancestry.

Humans, being vertebrates, have long wondered what sort of creature the first vertebrate was. A good candidate is a very primitive animal of some 400 million years ago called a sea-squirt—not the adult, but the larva. The adult is a highly specialized, immobile animal that fixes itself on the floor of the sea. The larva, however, is a free-swimming creature that has a small notochord (a primitive spinal cord, similar to what we grow as an embryo) that it loses before adulthood. A mutation that retarded the development of the sea-squirt could have been the start of all vertebrates.

Another question regarding our ancestors is this: Humans and chimpanzees have almost all of their genes in common. Why, then, do they differ so in both appearance and behavior?

Mutations in regulatory genes shed light on this question. Humans did not evolve from chimps, as noted earlier, but we and they had a common ape-like ancestor about five million years ago. Young chimps have a head shape that is much like that of a human. In later life, the head of the chimp or ape, with its low forehead and thrust-out jaw, departs considerably from the human's more vertical face and small jaws and teeth. The position of the spinal cord opening in the human skull is appropriate for erect posture, and is similar to that of embryonic apes and different from that of adult apes.

A regulatory mutation might have caused retarded development of the chimp/human ancestor, leading to the much longer period of dependency for the newborn human. During this childhood period, the brain continues to grow, and learning plays a very important part. So the young human comes to rely more on learning, rather than reacting instinctively. An additional benefit: family life is encouraged by the prolonged dependency on parents.

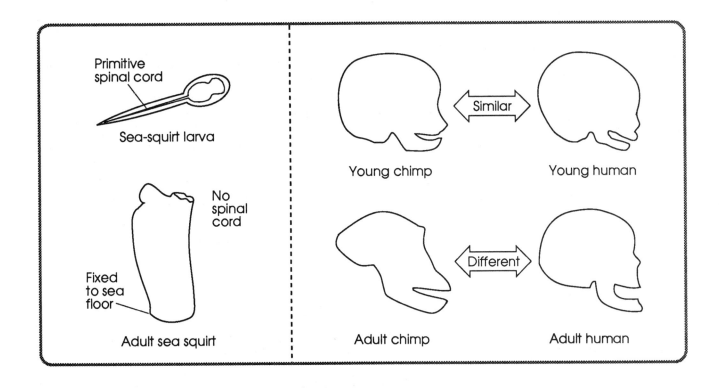

Primitive spinal cord

Sea-squirt larva

No spinal cord

Fixed to sea floor

Adult sea squirt

Young chimp   Similar   Young human

Adult chimp   Different   Adult human

Civilization arose from a different kind of evolution— cultural evolution—...

Our genes are the result of a slow evolution over billions of years. We also have another inheritance— our cultural inheritance—in which knowledge is passed from one individual to another. This kind of evolution is what is responsible for modern civilization: our buildings, roads, communications, universities and other institutions, including government—our way of life.

When humans began making stone tools, it was in some ways like an animal evolving sharper teeth or claws. In both cases, the improvement was valuable, and spread because its owner was more successful in surviving and producing more offspring.

Cultural evolution was, and is, much faster than natural evolution, and now completely overshadows it for us humans.

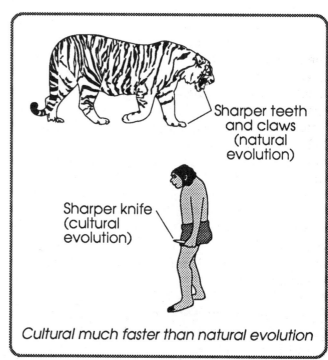

Sharper teeth and claws (natural evolution)

Sharper knife (cultural evolution)

Cultural much faster than natural evolution

which began when our ancestors stood erect, . . .

After our ancestors split off from the apes five million years ago they began walking and standing on their hind legs more and more, with vital consequences for us. Their upright posture freed their hands, which could then be used for carrying things, grasping rocks for use as tools, and other purposes. This in turn affected their brain, which—with the help of the prolonged period of dependency mentioned earlier—became larger and better at controlling their hands. The use of hands also freed their mouth and jaw; other animals use these for slashing, ripping and carrying. The new freedom of their mouth led to better vocal equipment, then to that most powerful of all tools, speech. This again affected their brain, making it larger and more capable of controlling language.

So standing up straight started a continuing interaction of hands, mouth and brain that vastly expanded the capabilities of primitive humans.

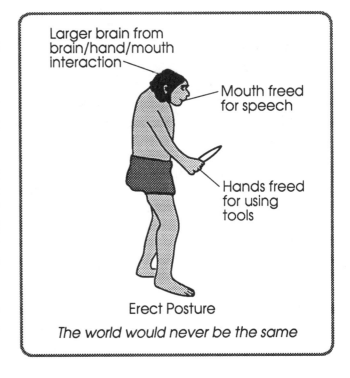

Larger brain from brain/hand/mouth interaction

Mouth freed for speech

Hands freed for using tools

Erect Posture

The world would never be the same

and, like genes, has been passed on for generations.

The effects of cultural evolution are so pervasive that they are difficult to appreciate. When we do simple, common things such as use a recipe, sing a song, play a game, get a thought from a book, we are often using ideas handed down by many unknown people for generations. Biologist Richard Dawkins suggests the word *memes* for concepts, tunes, sayings, etc., which, like genes, replicate and pass from brain to brain through generations. Memes that "work" survive and propagate. They're modified with time, of course. We build roads and bridges differently than we did a hundred years ago, but we still adapt for our own use ideas of that time.

Daily we employ a host of modern tools, instruments, and techniques. In doing so, we're not only "standing on the shoulders of giants" (in Newton's words), but also standing on the shoulders of many "little" people of previous generations who did their part and left the scene long ago.

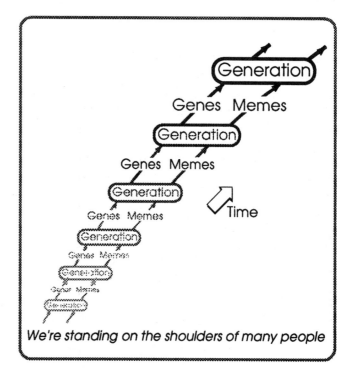

We're standing on the shoulders of many people

Without cultural evolution, we'd still be living in caves.

All humans alive today, as noted earlier, are members of the same species and have the same genes. The relationship is even broader than that: human genes have not changed for 40,000 years.

The differences in life-style between then and now could hardly be greater. People of that time lived in caves and had none of the trappings of modern society: buildings and houses, cars, roads and bridges, schools, radio and television, telephones, books, newspapers and magazines, stoves and refrigerators, and so on and on. But these differences would not have come about without cultural evolution: 40,000 years is too short a time for natural evolution to accomplish much, so if that were the only means for change, we'd still be living in caves.

40,000 years ago

Same genes, different environment (culture)

Today

Cultural evolution: a difference in life-styles

5

QUESTIONS AND UNCOMMON-SENSE ANSWERS

Sit down before fact as a little child, be prepared to give up every preconceived notion, follow humbly wherever and to whatever abysses nature leads, or you shall learn nothing.

— Thomas Henry Huxley

Questions that have troubled humanity for ages . . .

THERE'S NOTHING NEW about the questions we've been considering—they've troubled humankind for centuries. What *is* new is that we can begin to make intelligent judgments for the first time. The last forty years have unearthed 3 1/2-billion-year-old fossils, revealed the nature of the gene (DNA), and provided laboratory demonstrations that life's building blocks can arise from inanimate materials under primitive-earth conditions. The relatively new science of molecular biology has probed life's innermost secrets, exposing an underlying unity and tracing our roots to simple one-celled life.

Now that we can finally begin to answer the perennial questions, perhaps we first need to ask another question: Are we really ready for the answers— answers that point to a past so closely shared with other forms of life?

What is the origin of

THE UNIVERSE

LIFE

HUMAN BEINGS

?

Are we ready for the answers?

can now be answered, . . .

For the first time in history, we're close to understanding the origin of the universe, of life, and of human beings. In broad outline, the universe began some 15 – 20 billion years ago as a speck of energy, which started expanding and is still expanding. The simplest elements, hydrogen and helium, appeared first, forming into stars in which heavier atoms were created. Some stars later exploded as violent supernovas, scattering the heavier elements. In time these elements formed other "second generation" star systems, including our own solar system and earth. With the help of energy sources such as ultraviolet rays and lightning, primitive building blocks of life appeared on the early earth, from which the first living cell eventually emerged 3 1/2 to 4 billion years ago. This cell started to divide, and life has been evolving from it ever since.

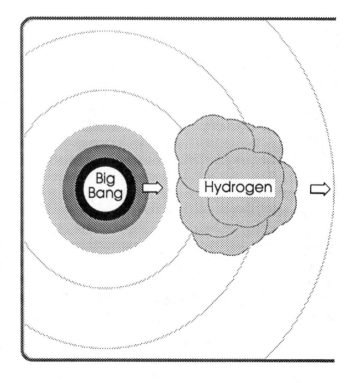

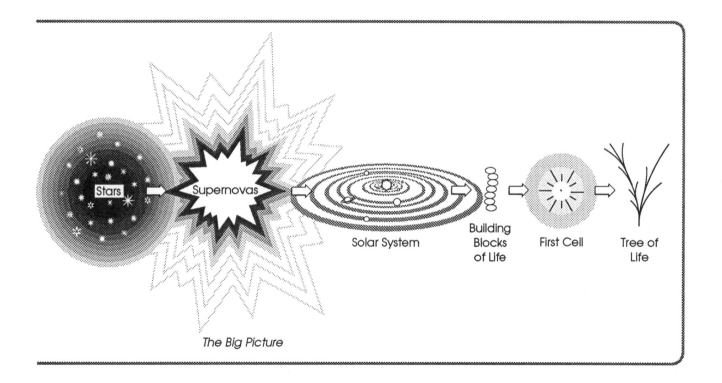

The Big Picture

but our common sense gets in the way.

Our "common sense" tends to get in the way of understanding. In a recent TV program, the speaker ridiculed the idea that "people came from hydrogen—a colorless, odorless, invisible gas." Their common sense outraged, the audience tittered appreciatively.

Hydrogen is more than just a gas: as the simplest element, with a single proton and electron (from which, along with neutrons, all elements are made), it serves as a building block for other elements. This building goes on right now in stars.

The problem with common sense was recognized by Descartes in 1637: "... everyone thinks himself so abundantly provided with it, that those even who are most difficult to satisfy in everything else, do not usually desire a larger measure of this quality than they already possess."

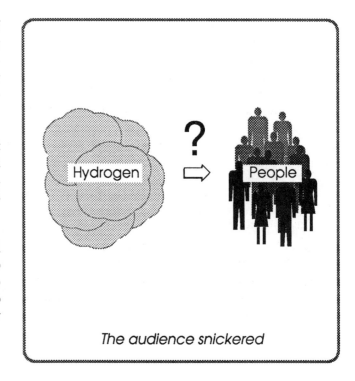

The audience snickered

There's another world out there, . . .

We live in a world of easily visible things, unaware of what's going on in the microscopic world. Early in this century, new ideas of matter and the atom changed our views about solid-looking objects like tables and chairs. There was, as physicist Ernest Rutherford put it, "a dissolution of all that is most solid into tiny specks floating in a void." Another instance of the growing attention "specks" have been getting in this century.

Our common sense is again being severely tested by molecular biology and related sciences. We must learn not to trust our common sense if we're to get anywhere with the kinds of questions we've been considering. And our imagination should not be limited to what we can see with the naked eye. With the poet William Blake, we can try "to see a world in a grain of sand."

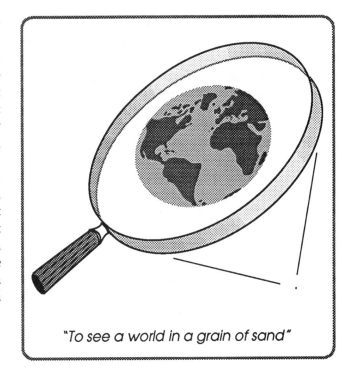

"To see a world in a grain of sand"

the microscopic world, . . .

A grain of sand seems small, but it's large compared to an atom. A single drop of water contains a million million billion (10^{21}) atoms—roughly the number of stars in the universe. If each atom in the drop of water were enlarged to the size of a grain of sand, the sand would fill a "beach" 500 feet wide, 20 feet deep, and 10,000 miles long!

With all those atoms crowded together in that drop of water, common sense says there couldn't be much space left over. But most of the drop *is* empty space. In a neutron star, this space has been crushed out by tremendous pressure. The density of such a star is so enormous that a tiny, just-visible speck (there's that speck again) would weigh *a million tons*!

Common sense? What we need more of for questions like these is *un*common sense.

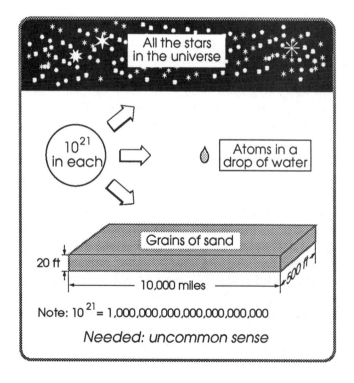

All the stars in the universe

10^{21} in each

Atoms in a drop of water

Grains of sand

20 ft

10,000 miles

500 ft

Note: 10^{21} = 1,000,000,000,000,000,000,000

Needed: uncommon sense

that we've been missing.

Microscopic life is everywhere. Our bodies swarm with bacteria. We tend to forget the bacteria and fungi in the air until we see a slice of bread molding or a fallen fruit rotting. Biologist Richard Dawkins describes how "it is raining DNA outside." A willow tree in his garden produces many small white flecks that carpet the ground—flecks that contain DNA. The flecks are composed mainly of bulky cellulose, which dwarfs the tiny DNA inside. Why, then, say it's raining DNA? Why not say it's raining cellulose? Well, the cellulose is what meets the eye, but it's just window dressing that will eventually disappear. It's the DNA that counts. It contains instructions for making willow trees that will in turn make more fluffy specks with DNA containing instructions for making ... And so on.

Just as future willow trees depend on these specks, future generations of humans depend on the genes that we presently carry in our bodies.

"It's raining DNA"

Some of our older beliefs are being revised, . . .

We've always wanted answers, so it's not surprising that we readily accepted rather far-out answers to troublesome questions in the past. At one time thunder was believed to be the work of the Greek god Zeus. People who pooh-pooh this idea may still think thunder causes milk to sour. Or they may believe the weather at the time of conception will have a strong effect on the forthcoming child. And not just the weather but the age, health, even the state of mind of the parents. There are primitive people today who see no connection between sexual intercourse and offspring. Not long ago common sense said the earth was flat. The idea that the earth was round and rotated was dismissed as absurd; obviously, everyone would fly off into space.

While we still have trouble predicting weather, we now have a better explanation for thunder ... and many other things.

It seemed like a good idea at the time

but there are now more miracles than ever, . . .

That some of our questions can now be answered doesn't make it seem any the less miraculous. Quite the contrary. We've found that a tiny speck can become (1) a human being, (2) all life on earth, or (3) the entire universe. And this strange story emerges *not* from science fiction but from dignified, sober, hard-headed science!

Certainly such specks—which must be the ultimate in low profiles, the "small is beautiful" concept carried to extreme—should lead us to question our more orthodox preconceptions and our much-lauded common sense, and look a little deeper. They teach us a little respect for specks, you might say. The real world is turning out to be much stranger and more remarkable than our limited imaginations could envision.

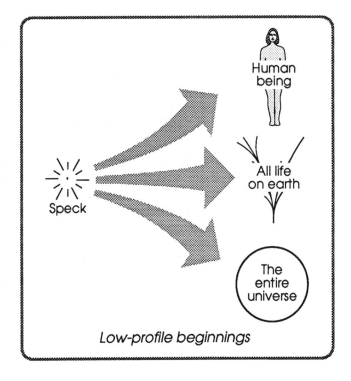

Low-profile beginnings

which reveal the pervasive unity of life.

We now have a new way of looking at the world, a new perspective. All life on earth, from the tiniest bacteria to ourselves, is made from atoms that originated in the stars, billions of miles away and billions of years ago. So ultimately, we're all made of stardust. But the unity of life goes much further than that—after all, the same could be said of tin cans, TV sets, and concrete buildings. There is a deep kinship of all creatures, living and extinct. We all came from that first pioneer cell some 3 1/2 billion years ago. Our genes have come down through the ages—modified, to be sure—from the earliest cells that struggled for life in the primeval sea. Our bodies preserve a record of these roots.

Astronomers have shown us we're not the center of the universe, as our ancestors believed. Our sun, one of billions of suns in the Milky Way galaxy, is not unusual. Nor is our galaxy, one of billions of galaxies, extraordinary in any way. In the latest assault on our ego, we find that we're just a twig on the tree of life, evolving with other creatures in an intricate growth pattern rooted in bacteria-like organisms and driven rather inelegantly with the help of copying errors. But this should not disturb us. It should teach us a new humility and appreciation of life, a respect and love for all creatures. After all, they are our relatives.

Though we share a common origin with other animals, a vast gulf has been opened up between us and them by our larger brain size and cultural evolution. The power of this newer evolution makes us realize what possibilities the future may hold for the illiterate and destitute people of the world. Inherently, they're like the rest of us—they have the same genes. What they need is a chance to fulfill the potential in those genes.

Can all our questions be answered? No. And even if they could, in science answers are never final. The origin of life is still a problem. It now seems solvable—molecular biology, still a young science, has made outstanding progress—but who knows? And we don't know and may never know what, if anything, came before the Big Bang. A series of expanding and collapsing universes? So there are still big gaps in our knowledge. Trying to fill them in will continue to stretch our imagination.

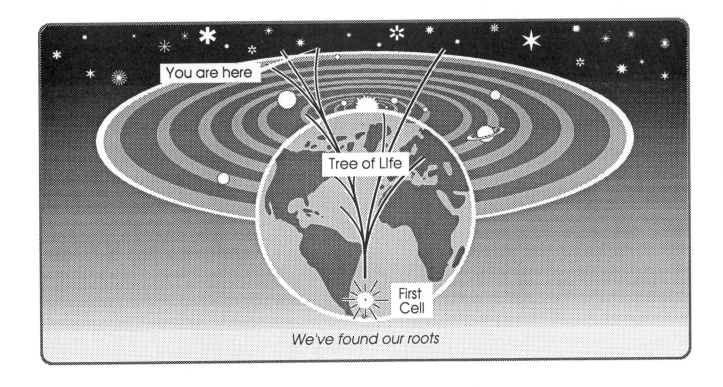

BIBLIOGRAPHY

The references below are suggested for further reading on the topics discussed in this book. Readers who would like only a few references that cover many of the ideas might begin with the following: Gamlin & Vines (Eds.), *The Evolution of Life*; Margulis & Sagan, *Microcosmos*; and Scheinfeld, *Your Heredity and Environment*.

Asimov, I. *The Wellsprings of Life*. Mentor Book, 1960.
———— *The Genetic Code*. Orion Press, 1962.
———— "The Inevitability of Life," *Of Matters Great and Small*. Doubleday & Co., 1975.
———— *Extraterrestrial Civilizations*. Crown Publishers, 1979.
Attenborough, D. *Life on Earth*. Little, Brown & Co., 1979.
Ayala, F.J. "The Mechanisms of Evolution," *Scientific American*, September 1978.
Bernal, J.D. *The Origin of Life*. World Publishing Co., 1967.
Blake, W. "Auguries of Innocence," *Harvard Classics*. P.F. Collier, 1910.
Borek, E. *The Sculpture of Life*. Columbia University Press, 1973.
Brachet, J. "The Living Cell," *Scientific American*, September 1961 (and other articles in this issue).
Cairns-Smith, A.G. *Genetic Takeover*. Cambridge University Press, 1982.
———— *Seven Clues to the Origin of Life*. Cambridge University Press, 1985.
Calder, N. *The Life Game*. Viking Press, 1973.
———— *The Key to the Universe*. Viking Press, 1977.
———— *Timescale: An Atlas of the Fourth Dimension*. Viking Press, 1983.
Carson, H.L. *Heredity and Human Life*. Columbia University Press, 1963.
Chaisson, E. *Cosmic Dawn: The Origin of Matter and Life*. Atlantic Monthly Press, Little, Brown & Co., 1981.
Clark, R.W. *The Survival of Charles Darwin*. Random House, 1984.

Crick, F. *Life Itself: Its Origin and Nature.* Simon & Schuster, 1981.

Dawkins, R. *The Selfish Gene.* Oxford University Press, 1976.

—————— *The Blind Watchmaker.* W.W. Norton & Co., 1986.

Day, W. *Genesis on Planet Earth: The Search for Life's Beginnings.* House of Talos Publishers, 1979.

Descartes, R. "Discourse on Method," *Harvard Classics.* P.F. Collier, 1910.

Dickerson, R.E. "Chemical Evolution and the Origin of Life," *Scientific American,* September 1978.

Dodson, E.O. *A Textbook of Evolution.* W.B. Saunders & Co., 1952.

Dulbecco, R. *The Design of Life.* Yale University Press, 1987.

Eddington, A.S. *Nature of the Physical World.* Cambridge University Press, 1928.

Eigen, M., et al. "The Origin of Genetic Information," *Scientific American,* April 1981.

Evans, B. *The Natural History of Nonsense.* Vintage Books, 1958.

Fox, R.F. *Energy and the Evolution of Life.* W.H. Freeman & Co., 1988.

Gamlin, L. & Vines, G. (Eds.) *The Evolution of Life.* Oxford University Press, 1987.

Gardner, M. *The New Ambidextrous Universe.* W.H. Freeman & Co., 1990.

Gehring, W.J. "The Molecular Basis of Development," *Scientific American,* October 1985.

Gilbert, M.S. *Biography of the Unborn.* Hafner Publishing Co., 1963.

Gould, S.J. *Ontegeny and Phylogeny.* Harvard University Press, 1977.

—————— *Ever Since Darwin.* W.W. Norton, 1977.

—————— *The Panda's Thumb.* W.W. Norton, 1980.

—————— *Hen's Teeth and Horses Toes.* W.W. Norton, 1983.

Gribbin, J. *In Search of the Double Helix.* Bantam Books, 1985.

Halle, L.J. *Out of Chaos.* Houghton Mifflin, 1977.

Hardin, G. *Nature and Man's Fate.* Rinehart & Co., 1959.

Hawking, S.W. *A Brief History of Time.* Bantam Books, 1988.

Hoffman, J.G. *The Life and Death of Cells.* Doubleday & Co., 1957.

Hofstadter, D.R. & Dennett, D.C. *The Mind's I.* Basic Books, 1981.

Hotton III, N. *The Evidence of Evolution.* American Publishing Co., 1968.

Hutchinson, P. *Evolution Explained.* David & Charles, 1974.

Huxley, A.F. "How Far Will Darwin Take Us," in D.S. Bendall (Ed.), *Evolution from Molecules to Men.* Cambridge University Press, 1983.

Jastrow, R. *Until the Sun Dies.* W.W. Norton, 1977.

Judson, R.F. *The Eighth Day of Creation.* Simon & Schuster, 1979.

Kendrew, J.C. *The Thread of Life.* Harvard University Press, 1966.

Kenyon, H.D. & Steinman, G. *Biochemical Predestination.* McGraw-Hill, 1969.

Lewin, R. *The Thread of Life.* Smithsonian Books (W.W. Norton), 1982.

Luria, S.E. *Life: The Unfinished Experiment.* Chas. Scribner's Sons, 1973.

Margulis, L. & Sagan, D. *Microcosmos: Four Billion Years of Microbial Evolution.* Summit Books, Simon & Schuster, 1986.

Mayr, E. "Evolution," *Scientific American,* September 1978 (and other articles in this issue).

Monod, J. *Chance and Necessity,* A. Knopf, 1971.

Montagu, A. (Ed.) *Science and Creationism,* Oxford University Press, 1984.

Moore, R. *The Coil of Life,* A. Knopf, 1961.

Nillson, L. *A Child Is Born,* Delacorte Press, 1977.

Pfeiffer, J. et al. *The Cell.* Time, Inc. 1964.

Prigogine, I. & Stengers, I. *Order Out of Chaos.* Bantam Books, 1984.

Prusiner, S.B. "Prions," *Scientific American,* October 1984.

Rush, J.H. *The Dawn of Life.* Signet Science Library, 1957.

Sagan, C. *The Cosmic Connection.* Anchor Press, Doubleday & Co., 1973.

Scheinfeld, A. *Your Heredity and Environment.* J.B. Lippincott Co., 1965.

Schopf, J.W. "The Evolution of the Earliest Cells," *Scientific American,* September 1978.

Shapiro, R. *Origins: A Skeptic's Guide to the Creation of Life on Earth.* Summit Books, Simon & Schuster, 1986.

Silk, J. *The Big Bang.* W.H. Freeman & Co., 1980.

Simpson, G.G. *Biology and Man.* Harcourt Brace & World, 1969.

Sinnott, E.W. *The Bridge of Life.* Simon & Schuster, 1966.

Snively, W.D. *Sea of Life.* David McKay Co., 1969.

Stebbins, G.L. *Darwin to DNA, Molecules to Humanity.* W.H. Freeman & Co., 1982.

Tanner, J. et al. *Growth.* Time, Inc., 1965.

Valentine, J.W. "The Evolution of Multicellular Plants and Animals," *Scientific American,* September 1978.

Watson, J.D. et al. *Recombinant DNA: A Short Course*. W.H. Freeman & Co., 1983.

———— et al. *Molecular Biology of the Gene*. Benjamin/Cummins Publishing Co. (4th ed.), 1987.

Weinberg, R.A. "The Molecules of Life," *Scientific American*, October 1985 (and other articles in this issue).

Weintraub, P. (Ed.) *The OMNI Interviews*. OMNI Press, 1984. (Interviews with C. Ponnamperuma, E. Mayr, I. Prigogine.)

Wilson, A.C. "The Molecular Basis of Evolution," *Scientific American*, October 1985.

Wilson, E.O. et al. *Life on Earth*. Sinauer Assoc., 1973.

Winchester, A.M. *Heredity and Your Life*. Dover Publications (2nd ed.), 1960.

Woese, C.R. "The Primary Lines of Descent and the Universal Ancestor," in D.S. Bendall (Ed.), *Evolution from Molecules to Men*. Cambridge University Press, 1983.

Wooldridge, D.E. *The Machinery of Life*. McGraw-Hill, 1966.

Woosley, S. & Weaver, T. "The Great Supernova of 1987," *Scientific American*, August 1989.

GLOSSARY

Many of the following terms can be described in different ways depending on the technical level at which they are being discussed. The meanings given below are at about the level used in this book.

amino acids: building blocks of protein; there are 20 different types of amino acids.

ATP (adenosine triphosphate): a molecule that is used universally in cells to store energy; energy from respiration and fermentation is stored in ATP.

autocatalysis: self-catalysis, in which a chemical reaction is catalyzed by one of its own products.

bacteria: microscopic organisms that (along with viruses) have existed longer than any other life form; consist of a single cell without a nucleus.

bacterial cell: simple cell without a nucleus, as distinguished from nucleated cells.

base pair: a step of the DNA spiral staircase; the only possible pairs are A-T (adenine-thymine) or G-C (guanine-cytosine). There are about three billion base pairs in the genes of a single human cell.

bases: see nucleic acid bases.

Big Bang: the beginning of time, a singularity that represented the origin of the universe; our theory of gravity extends back to the "Planck time," or 10^{-43} seconds after the singularity, at which time "all the matter we now see in the universe, comprising some millions of galaxies, was compressed within a sphere of radius equal to one-thousandth of a centimeter, the size of the point of a needle" (Joseph Silk).

body cell: common cells of which our bodies are composed; contrasted with germ cells, used for reproduction. A typical number of body cells in an adult human is 60 trillion (J.G. Hoffman), but one finds estimates varying from 10 – 100 trillion. Cells vary in size: tissue cells are from 0.0002 to 0.0008 inches in diameter, and nerve cells may be as long as four feet.

Cambrian explosion: the period beginning about 570 million years ago when many hard animal body parts were left as fossils in rocks world-wide, indicating a great variety of life at that time.

catalysis: acceleration of a chemical reaction by a substance ("catalyst') that is not changed by the reaction.

cell: the basic unit of life, of which there are two main types: the non-nucleated cells of bacteria and the nucleated cells of all multicellular organisms.

cell nucleus: cell structure containing the chromosomes and genes of nucleated cells.

chloroplasts: bacterialike cell entities that contain chlorophyll, which is used in photosynthesis by plants and some bacteria; like mitochondria, now believed to be relics of ancient bacteria that combined symbiotically with nucleated cells.

chromosomes: threadlike bodies in the cell nucleus that contain the genes in a linear array; body cells have 23 pairs of chromosomes and germ cells only 23 chromosomes.

chromosome shuffling: in forming germ cells, one chromosome is selected at random from each pair of the mother's chromosomes and one from each pair of the father's chromosomes; a type of "recombination."

chordate: a broad biological category (a "phylum") that includes vertebrates—fish, amphibians, reptiles, birds, and mammals—as well as some "lower chordates" that are not vertebrates, such as amphioxus and tunicates. All chordates have, at some stage in their lives, a notochord (a spine of cartilage), a nerve cord above this, and gill slits just behind the mouth. In vertebrates, the notochord is replaced by a bony backbone with separate vertebrae.

conception: the union of the father's sperm and mother's egg to produce a fertilized egg.

deoxyribose sugar: the type of sugar which, together with a phosphate and base, forms a nucleotide—a DNA building block; has one less oxygen molecule than ribose sugar.

DNA: deoxyribonucleic acid, the substance genes are made of; through the genetic code, it directs the building of proteins.

double helix: the spiral-staircase structure of the DNA molecule; "steps" are bases, "arms" are phosphates and sugars.

egg cell: the female germ cell, released about once a month from the ovaries, starting at puberty; contains half the number of chromosomes of a body cell.

embryo: the newly conceived offspring through the eighth week after conception.

enzymes: protein molecules in the cell that act as catalysts in bringing together other molecules and speeding up their reaction time by a factor of a million or more; their effectiveness depends on the unique binding site formed when the protein molecule folds.

fallopian tubes: a pair of female tubular channels connecting the ovaries and womb, through which the fertilized egg passes after conception.

fermentation: the anaerobic ("without oxygen") breakdown of organic molecules such as sugar to produce energy plus waste products of acids and alcohols. (Sometimes called "anaerobic respiration.")

fetus: the offspring prior to birth, from end of eighth week after conception until birth.

fungi: with animals and plants, one of the three main branches of nucleated-cell organisms; develop from spores.

genes: chromosome elements that are the carriers of inherited characteristics; each gene consists of a portion of a DNA molecule whose beginning and end are defined by base triplets (see **genetic code,** below).

genetic code: a code that translates a triplet of DNA bases into a specific amino acid; a string of amino acids can form a protein, but in humans only 1% – 2% of DNA actually "codes for" protein.

germ cells: cells used for reproduction—egg cells for females, sperm for males; different from common body cells in that they have only half the normal number of chromosomes (the term "half cells" was suggested by A.M. Winchester). When united (egg and sperm), the resulting fertilized egg has the full set of chromosomes. The egg cell is about 0.005 inches in diameter; a sperm has a long tail, about 0.0024 inches.

hominids: man-ape animals that walk erect with their hands hanging free; we humans are now the only hominids (family *Hominidae*), with the apes being in a separate family (*Pongidae*).

Homo sapiens: the human species—genus *Homo*, species *sapiens*, literally "wise man"—extending back some 200,000 years; modern humans from about 40,000 years ago to the present are sometimes called *Homo sapiens sapiens* to distinguish them from earlier humans.

isomer: a chemical compound similar to another in number of atoms of the same elements, but differing in structural arrangement. Of particular interest in life studies is stereoisomerism, which has to do with symmetry, or "handedness." In living things, all amino acids are left handed, and all nucleic acid sugars are right handed.

jumping genes: segments of DNA (called "transposons") that are able to move from chromosome to chromosome apparently at random; first suggested by biologist Barbara McClintock in the 1940s and finally accepted in the 1970s.

mammal: a broad biological category that includes vertebrate animals that nurse their young with milk via mammary glands; we humans are mammals.

meme: the word suggested by biologist Richard Dawkins for ideas, customs, knowledge, and institutions that are passed on from generation to generation, not biologically by genes but by teaching and other communication. Amram Scheinfeld uses the term *social genes* for the same concept.

methane bacteria: a form of anaerobic ("without oxygen") bacteria that can use carbon dioxide and hydrogen, or organic acids, to produce energy and methane gas.

mitochondria: bacterialike cell entities that oxidize food molecules to produce energy that is stored in ATP molecules, with carbon dioxide and water as waste products; as suggested in 1918 and more recently advocated by biologist Lynn Margulis, these (along with chloroplasts for plants) are now widely believed to be relics of ancient bacteria that combined symbiotically with nucleated cells.

mutation: changes in the DNA structure, most of which are either harmful or neutral; caused by random copying errors, or by radiation or chemicals.

natural selection: weeding out of the less adaptable organisms by the natural environment, leaving the more adaptable to survive and propagate; compared by Darwin to artificial selection, in which, for example, dogs are selected by humans for certain characteristics (all dogs are one species, descended from the wolf).

neutron star: a tiny star (typically only a mile in diameter) remaining after tremendous pressures have crushed it to nothing but protons and neutrons (the nuclei of atoms); "so dense that a speck of it—just barely visible—would weigh a million tons" (Carl Sagan).

nucleated cells: cells having a nucleus containing chromosomes and genes; all multicelled organisms are composed of such cells.

nucleic acid bases: constituents of nucleic acids that make up the "steps" of DNA and RNA molecules; four types occur in DNA (adenine, cytosine, guanine, and thymine—usually abbreviated as A, C, G, and T, respectively), and four in RNA (the same first three as DNA, with uracil, U, replacing thymine).

nucleotide: a constituent of DNA and RNA, consisting of a base attached to a sugar-phosphate unit.

organic molecules: complex carbon-containing molecules, found in all living things. There are four main types: carbohydrates (which include sugars, starches, cellulose, etc.), lipids, amino acids, and nucleotides. Before life existed there were organic molecules in the ocean formed by the same processes that eventually formed life itself.

ovaries: female sex glands, from which eggs are released about once a month.

photosynthesis: the use of water, carbon dioxide and sunlight to make sugars, emitting oxygen as a waste product; both plants and certain bacteria can do this.

placenta: female organ that keeps the developing embryo/fetus firmly attached to the womb; grows from the fertilized egg (and is therefore not a part of the mother), and connects to the umbilical cord, which transfers nutrients and oxygen from mother to embryo and waste products the other way.

primate: a broad biological category (an "order") of mammals that includes humans, apes, and monkeys, as well as marmosets and lemurs.

progenote: single-celled ancestor of all cells, named by Carl R. Woese; led to several lines of single-cell descent, and ultimately to all life on earth.

protein: a basic constituent of the body, forming both structural material as different kinds of cells, and enzymes; made up of long strings of 20 different kinds of amino acids.

respiration: the use of oxygen to break down organic molecules to produce energy, carbon dioxide and water. (Respiration in this book refers to aerobic respiration—the use of oxygen. Fermentation is sometimes referred to as anaerobic respiration—without oxygen.)

ribosomes: extremely small and numerous cell particles where amino acids are joined with the help of enzymes, and as prescribed by RNA, to make proteins; so-named because of their RNA activity.

ribose sugar: the type of sugar which, together with a phosphate and base, forms a nucleotide—an RNA building block.

RNA: ribonucleic acid, a substance that conveys the DNA information to other parts of the cell outside the nucleus, particularly to the ribosomes, where proteins are made.

sperm: the male germ cell, generated in the testes continually starting with puberty, and containing half the number of chromosomes of a body cell.

stromatolites: fossilized mats of ancient bacterial origin that still exist near the seashore in certain areas; the oldest are about 3 1/2 billion years old.

symbiosis: the living together of members of two or more species; examples are cows, which have cellulose-digesting bacteria in their gut; termites, which are similarly endowed; and plants that depend on nitrogen-fixing bacteria in their roots.

umbilical cord: a cord connecting the placenta and embryo/fetus, through which food and oxygen passes from the mother and waste from the embryo/fetus.

vertebrate: a broad biological category (a "subphylum" of chordates) that includes fish, amphibians, reptiles, birds, and mammals, all of which have a segmented backbone with separate vertebrae.

viroid: even smaller and simpler than a virus, it consists of a molecule of RNA without a protective coat; can cause disease in plants.

virus: a miniscule form of matter (diameter 10^{-6} cm, or about 0.0000004 in.) that is on the threshold between life and non-life and consisting of a protective coat of protein and a molecule of nucleic acid (DNA or RNA); may be found in the form of "lifeless" crystals, and can reproduce only within the cells of animals, plants, or bacteria.

womb: a female organ (uterus) for containing and nourishing the embryo/fetus prior to birth.

X,Y chromosomes: chromosomes in the 23rd pair that determine sex; XX is female, XY is male.

INDEX

ABOUT THE AUTHOR

W. J. ("Jim") Howard is a writer and part-time consultant in systems analysis and mathematics. Formerly a mathematician with the Rand Corporation and Planning Research Corporation, he has done consulting work for General Electric, TRW, Aerospace Corporation, Matrix Corporation, Serendipity Associates, Xerox Corporation, Xyzyx Information Corporation and others. He now specializes in making technical information understandable to the general reader. To this end he has produced a variety of books, usually under contract. Some examples: *A Simple Manual on Queues*, *Vegetable Gardening*, and *Kitchen and Bathroom Remodeling* (Xyzyx Information Corporation—Dr. Kay Inaba); *Venereal Disease*, co-authored with Dr. Ruth Schlesinger (USC School of Medicine—Dr. Art Ulene); *Introduction to Interactive Accounting System* and *Turn Around Time: A Key Factor in Printing Productivity* (Xerox Corporation).

The present book was typeset and illustrated by the author (except for parts of two illustrations, which are from the Kwikee® Potpourri collection), with a lot of help from an Apple Macintosh® computer and LaserWriter® printer, and Aldus PageMaker® and Adobe Illustrator™ software.

Order Form

Coast Publishing
P.O. Box 3399
Coos Bay, OR 97420

Please send me _____ copies of LIFE'S BEGINNINGS @ $9.95 per copy.

No charge for postage and handling. Books will be sent by priority mail (first class).

Total enclosed: $_____
Please send check or money order, U.S. dollars.

Name:_____

Address:_____

City:_____State:_____Zip:_____